Advances in

BOTANICAL RESEARCH

incorporating *Advances in Plant Pathology*

VOLUME 21

Advances in

BOTANICAL RESEARCH

incorporating *Advances in Plant Pathology*

Editor-in-Chief

J. A. CALLOW · *School of Biological Sciences, University of Birmingham, Birmingham, UK*

Editorial Board

Advances in

BOTANICAL RESEARCH

incorporating *Advances in Plant Pathology*

Series editor

J. A. CALLOW

School of Biological Sciences
University of Birmingham
Birmingham, UK

VOLUME 21

Edited by

John H. Andrews

Department of Plant Pathology,
University of Wisconsin-Madison
Madison, USA

Inez C. Tommerup

Division of Forestry, CSIRO
Perth, Australia

1995

ACADEMIC PRESS
Harcourt Brace & Company, Publishers
London   San Diego   New York   Boston
Sydney   Tokyo   Toronto

This book is printed on acid-free paper

ACADEMIC PRESS LIMITED
24/28 Oval Road,
London NW1 7DX

United States Edition published by
ACADEMIC PRESS INC.
San Diego, CA 92101

A catalogue record for this book is available from the British Library
ISBN 0-12-005921-5

Transferred to digital printing 2006

Typeset by Colset Private Limited, Singapore
Printed and bound by CPI Antony Rowe, Eastbourne

CONTENTS

Defense Responses of Plants to Pathogens

ERICH KOMBRINK and IMRE E. SOMSSICH

On the Nature and Genetic Basis for Resistance and Tolerance to Fungal Wilt Diseases of Plants

C. H. BECKMAN and E. M. ROBERTS

Implication of Population Pressure on Agriculture and Ecosystems

ANNE H. EHRLICH

Plant Virus Infection: Another Point of View

G. A. DE ZOETEN

The Pathogens and Pests of Chestnuts

SANDRA L. ANAGNOSTAKIS

Fungal Avirulence Genes and Plant Resistance Genes: Unraveling the Molecular Basis of Gene-for-gene Interactions

PIERRE J. G. M. DE WIT

Phytoplasmas: Can Phylogeny Provide the Means to Understand Pathogenicity

BRUCE C. KIRKPATRICK and CHRISTINE D. SMART

Use of Categorical Information and Correspondence Analysis in Plant Disease Epidemiology

S. SAVARY, L. V. MADDEN, J. C. ZADOKS and H. W. KLEIN-GEBBINCK

CONTRIBUTORS TO VOLUME 21

S. L. ANAGNOSTAKIS, *The Connecticut Agricultural Experiment Station, New Haven, Connecticut 06504-1106, USA*

C. H. BECKMAN, *Department of Plant Sciences, University of Rhode Island, Kingston, Rhode Island 02881, USA*

P. J. G. M. DE WIT, *Department of Phytopathology, Wageningen Agricultural University, PO Box 8025, 6700EE Wageningen, The Netherlands*

G. A. DE ZOETEN, *Department of Botany and Plant Pathology, Michigan State University, East Lansing, Michigan 48824, USA*

A. H. EHRLICH, *Department of Biological Sciences, Stanford University, Stanford, California 94305, USA*

B. C. KIRKPATRICK, *Department of Plant Pathology, University of California, Davis, California 95616, USA*

H. W. KLEIN-GEBBINCK, *IRRI, Department of Plant Pathology, PO Box 933, 1099 Manila, Philippines (Present address: Department of Plant Science, University of Alberta, Edmonton, Alberta T6G 2P5, Canada)*

E. KOMBRINK, *Max-Planck-Institut für Züchtungsforschung, Abteilung Biochemie, Carl-von-Linné-Weg 10, D-50829 Köln, Germany*

L. V. MADDEN, *Department of Plant Pathology, Ohio Agricultural Research and Development Center, The Ohio State University, Wooster, Ohio 44691-4096, USA*

E. M. ROBERTS, *Department of Plant Sciences, University of Rhode Island, Kingston, Rhode Island 02881, USA*

S. SAVARY, *ORSTOM Visiting Scientist at International Rice Research Institute, Entomology and Plant Pathology Division, PO Box 933, 1099 Manila, Philippines*

C. D. SMART, *Department of Plant Pathology, University of California, Davis, California 95616, USA*

I. E. SOMSSICH, *Max-Planck-Institut für Züchtungsforschung, Abteilung Biochemie, Carl-von-Linné-Weg 10, D-50829 Köln, Germany*

J. C. ZADOKS, *Department of Phytopathology, Wageningen Agricultural University, PO Box 8025, 6700EE Wageningen, The Netherlands*

SERIES PREFACE

Advances in Botanical Research is one of Academic Press' longest standing serials, and has established an excellent reputation over more than 30 years. *Advances in Plant Pathology*, although somewhat younger, has also succeeded in attracting a highly respected name for itself over a period of more than a decade.

The decision has now been made to bring the two serials together under the title of *Advances in Botanical Research incorporating Advances in Plant Pathology*. The resulting synergy of the merging of these two serials is intended to greatly benefit the plant science community by providing a more comprehensive resource under one 'roof'.

John Andrews and Inez Tommerup, the previous editors of *Advances in Plant Pathology*, are now on the editorial board of the new series. Our joint aim is to continue to include the very best articles, thereby maintaining the status of a high impact factor review series.

PREFACE

The following collection of articles represents issues of current and future interest in plant pathology. The essays range from a very basic to an applied focus (deliberately intermingled within the volume) on topics written for a wide audience, including undergraduate and postgraduate students, researchers, and teachers. Our goal has been both to draw insights from relevant biological disciplines into the realm of plant pathology and to reveal the general principles of plant pathology to the broad audience of biologists. The authors have been encouraged to write in a thought-provoking and, where appropriate, controversial manner.

One of the most active and rapidly breaking areas in plant pathology is the molecular basis for disease resistance, in essence how genes and gene products of the pathogen interact with those of the host. Several articles in the volume take up this theme, beginning with a general overview of defense responses by E. Kombrink and I. E. Somssich. In more specific treatments, C. H. Beckman and E. M. Roberts then assess the nature and genetic basis for resistance to fungal wilt diseases; G. A. de Zoeten advances a controversial intepretation of the early events in plant virus infection; and P. J. G. M. de Wit considers the interaction between fungal avirulence genes and plant resistance genes.

Though the immense and complex problems posed by an increasing world population are less heralded in the public media than was the case some two decades ago, they continue to increase. A. H. Ehrlich considers some of the impacts of increasing population pressure on agricultural and non-agricultureal ecosystems, including the loss in biodiversity. Pathogens as well as humans can have a devastating impact on ecosystems. There are few more striking examples of this than the elimination of chestnut as a significant component of the forest landscape in the Eastern United States by the introduction of the blight fungus *Cryphonectria parasitica*. S. L. Anagnostakis reviews the status of chestnuts and prospects for control of its various pests and diseases.

Likewise, the introduction of new techniques and analytical methods can have far-reaching consequences. The most obvious of these is the molecular revolution underway in numerous subdisciplines of biology and medicine. A good example is the use of gene sequencing and recombinant DNA technologies for inference of taxonomic and phylogenetic relationships, illustrated here for mycoplasma-like organisms by B. C. Kirkpatrick and C. D. Smart. Analogously, advances in statistics and computational sciences, though less widely recognized, have greatly facilitated experimental design and data analysis. S. Savary and colleagues show how plant pathologists have developed and adapted aspects of systems analysis, including correspondence analysis and categorical information, to the handling of complex data sets in epidemiology.

John H. Andrews
Inez C. Tommerup

Defense Responses of Plants to Pathogens

ERICH KOMBRINK and IMRE E. SOMSSICH

*Max-Planck-Institut für Züchtungsforschung, Abteilung Biochemie,
Carl-von-Linné-Weg 10, D-50829 Köln, Germany*

Advances in Botanical Research Vol. 21
incorporating Advances in Plant Pathology
ISBN 0–12–005921–5

I. INTRODUCTION

Plants constitute the largest and most important group of autotrophic life-forms on earth. Their abundant organic material serves as a nutritional source for all heterotrophic organisms, including animals, insects, and microbes. Despite the large number of fungi and bacteria involved in the decomposition of dead plant material, only very few of these potential pathogens have acquired the ability to colonize living plants by evolving from saprophytic organisms to perthotrophic or biotrophic pathogens. Thus, most plants exhibit natural resistance to microbial attack (*non-host resistance*, basic incompatibility), and therefore, disease is the exception rather than the rule. Non-host resistance is the consequence of either the inability of the parasite to recognize and to infect a plant or the ability of a plant to rapidly and successfully activate its defense mechanisms, invariably leading to protection and to the resistance phenotype. In contrast, relatively few specialized pathogens have evolved the means to successfully parasitize a plant host and establish basic compatibility by damaging or weakening the plant with toxins, by inhibiting host defense mechanisms, or by escaping recognition, and thus avoiding induction of host defense responses. The host range of a pathogen is usually very limited and often restricted to a single plant species. In such compatible interactions, the host plant may show severe disease symptoms which, in extreme cases, when agriculturally and economically important crop plants are affected, may have costly and even devastating consequences. The epidemic spread of the late blight fungus *Phytophthora infestans*, causing destruction of the staple crop potato (*Solanum tuberosum*) in many European countries and culminating in the Irish potato famine of 1846, is one such example (Schumann, 1991).

Selection and breeding for resistant cultivars and coevolution of pathogen races with altered virulence have led to the development of host–pathogen combinations in which only certain cultivars can be colonized by particular races of the pathogen, whereas others retain resistance (*cultivar resistance*). Several pathosystems that show this type of cultivar–race specificity, with gene-for-gene relationships of genetically defined resistance (*R*) genes of the host and corresponding avirulence (*avr*) genes of the pathogen, have been extensively studied for plant–fungus interactions as well as for interactions involving bacteria, nematodes, and insects. Such studies led to various models to explain the specificity underlying the determination of resistance (Keen, 1990; de Wit, 1992; Pryor and Ellis, 1993). The elicitor–receptor model is the most widely accepted one because it is consistent with most of the available genetic and biochemical data. It emphasizes the importance of the recognition process between host and pathogen and postulates that either primary avirulence gene products or metabolites resulting from their catalytic activity (elicitors) are recognized by specific plant target molecules (receptors) encoded by major disease resistance genes.

However, the precise biochemical mechanisms that determine the resistant phenotype are largely unknown, although specificity and resistance are often genetically controlled. Extensive studies of many different plant–pathogen interactions showed that, upon recognition of a pathogen, plants utilize a large arsenal of defense mechanisms to protect themselves. These molecular mechanisms appear to be very similar in non-host and cultivar resistance, despite plant species-specific differences, and they include preformed as well as induced defense mechanisms. The biochemical responses usually observed are the synthesis and deposition of phenolic compounds and proteins in the cell wall, rapid localized cell collapse and death (hypersensitive response, HR), accumulation of antimicrobial phytoalexins, and synthesis of pathogenesis-related proteins (PR proteins) comprising a wide range of different plant defense proteins, including chitinase and 1,3-β-glucanase. A number of excellent and comprehensive reviews has recently been published on the various biochemical and molecular responses of plants to pathogens as well as on important related aspects such as recognition, specificity and signal transduction (Bowles, 1990; Dixon and Harrison, 1990; Cutt and Klessig, 1992; Keen, 1992; Atkinson, 1993; Collinge et al., 1994; Dixon et al., 1994). In this chapter, it is not our intention to present a comprehensive compilation of the extensive literature on plant defense responses, but rather to provide an update of the more recent data and developments within this field of research. Following a description of selected pathogen virulence functions and active plant defense responses, we will mainly discuss signaling events and the regulation of plant defense genes, in particular their temporal and spatial expression patterns. Strategies for enhancing disease resistance by genetic engineering based on the knowledge obtained from the analysis of plant–pathogen interactions will be briefly addressed, whereas the aspects of specificity and recognition will be largely omitted. The preferential discussion of plant responses to fungal pathogens merely reflects our own research interests.

II. PATHOGEN INGRESS AND PLANT RESISTANCE

A. VIRULENCE FUNCTIONS OF PATHOGENS

Plant pathogens usually express several virulence functions that increase their ability to colonize and damage host plants (Fig. 1) (Salmond, 1994; Schäfer, 1994). Since virulence of the pathogen and resistance of the plant are reciprocal concepts, a discussion of plant defense responses necessarily entails a consideration of virulence functions of the pathogen. Some of these are general mechanisms, such as the production of enzymes, toxins, or plant growth regulators that damage or alter plant cells and provide an optimal environment for the pathogen. Others are highly specific, e.g. host-specific toxins, allowing pathogenesis only on certain cultivars or plant species and thus determining specificity of the interaction.

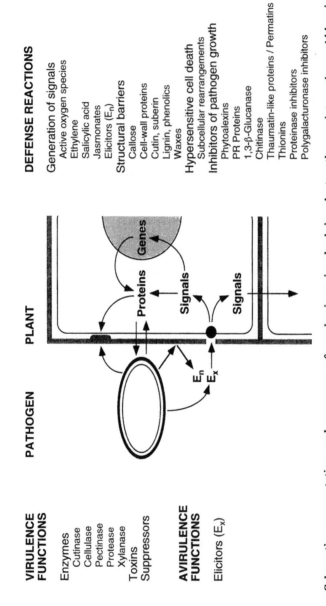

VIRULENCE FUNCTIONS

Enzymes
  Cutinase
  Cellulase
  Pectinase
  Protease
  Xylanase
Toxins
Suppressors

AVIRULENCE FUNCTIONS

Elicitors (E_x)

PATHOGEN   **PLANT**

E_n
E_x

Genes

Proteins

Signals

Signals

DEFENSE REACTIONS

Generation of signals
  Active oxygen species
  Ethylene
  Salicylic acid
  Jasmonates
  Elicitors (E_n)
Structural barriers
  Callose
  Cell-wall proteins
  Cutin, suberin
  Lignin, phenolics
  Waxes
Hypersensitive cell death
  Subcellular rearrangements
Inhibitors of pathogen growth
  Phytoalexins
  PR Proteins
  1,3-β-Glucanase
  Chitinase
  Thaumatin-like proteins / Permatins
  Thionins
  Proteinase inhibitors
  Polygalacturonase inhibitors

Fig. 1. Schematic representation and summary of mechanisms involved in plant–pathogen interactions. Abbreviations: E_n, endogenous elicitor; E_x, exogenous elicitor. Modified after Kombrink *et al* (1993); see text for details.

Before a pathogen can establish an infection, it must overcome the physical barriers protecting all plant cells from their environment, namely the cuticle and the cell wall. The cuticle consists of hydroxy fatty acids esterified with either phenolics (cutin) or primary alcohols (waxes) and the cell wall consists of polymers of carbohydrates (cellulose, hemicellulose, pectins), amino acids (hydroxyproline-rich glycoproteins, glycine-rich proteins, enzymes) and phenolics (lignin, phenolic esters). Cutin is found in the aerial parts of the plant, whereas the structurally related suberin, an insoluble phenolic matrix similar to lignin with fatty acids attached, is associated with wound healing and also found in barks and the underground parts of the plant. The importance of the cuticle, the outermost physical barrier, in plant disease resistance is presumably limited and controversial. Various chemicals present in cuticular waxes are inhibitory to the growth of pathogenic organisms. On the other hand, some cuticle components stimulate spore germination and germ tube development thus playing a positive role in pathogenesis (Kolattukudy, 1980). Many pathogens, particularly bacteria and viruses, depend on wounds or natural openings, such as stomata or hydathodes, to enter the plant. Others have developed sophisticated ways to physically penetrate the cuticle with an appressorium from which an infection peg extends into the plant. Still other fungi secrete extracellular enzymes, among which an esterase specific for cutin (cutinase) has been proposed as being crucial for infection of plants (Schäfer, 1994). Recently, strong evidence for the involvement of cutinases in plant–pathogen interactions was obtained by insertion of a cutinase gene, derived from the pea pathogen *Nectria haematococca* (anamorph *Fusarium solani* f. sp. *pisi*), into *Mycosphaerella* spp., a wound pathogen of papaya fruits. Transformants of this pathogen were now able to directly penetrate papaya fruit through the non-wounded cuticle (Dickman *et al.*, 1989). The importance of cutinase as a fungal virulence factor, however, has been questioned by construction of cutinase-deficient mutants of *Nectria haematococca* via transformation-mediated gene disruption of the single cutinase gene of a "highly" virulent strain, which nevertheless showed no difference in pathogenicity and virulence on pea compared to the wild type (Stahl and Schäfer, 1992). Conversely, overexpression of a heterologous (*Nectria haematococca*) cutinase in *Cochliobolus heterostrophus*, a pathogen of corn foliage, did not affect the virulence or specificity of the transformants on all organs of either its host (corn) or a non-host (pea) (Oeser and Yoder, 1994). The conclusion drawn from these latter results was that cutinase of *Nectria haematococca* is not essential for infection of pea. However, the role of cutinase for pathogenicity remains controversial since a careful reevaluation of the cutinase gene-disrupted *Nectria* mutant showed that its virulence was indeed impaired (Rogers *et al.*, 1994). Furthermore, it must be remembered that highly virulent strains do have multiple cutinase genes and also high levels of extracellular enzyme activity which could assist the pathogen to infect its host, although alternative routes into the tissue, i.e. via stomata, may also be used (Rogers *et al.*, 1994).

Once the cuticle is breached, the pathogen must still contend with the cell wall during its attempt to colonize a host. Fungal and bacterial pathogens secrete a large arsenal of hydrolytic enzymes capable of digesting cell wall polymers. These enzymes include, among others, cellulases, pectinases, xylanases and proteases (Salmond, 1994; Walton, 1994). The most compelling evidence for cell wall degrading enzymes being important virulence functions comes from genetic experiments with the soft-rot bacterium *Erwinia*. Directed deletions of pectate lyase genes resulted in a decreased virulence of the mutants on a plant host (Beaulieu *et al.*, 1993). However, some mutants, including one in which all structural pectate lyase genes of the multigene family were affected, retained pathogenicity and the capability of macerating plant tissue (Kelemu and Collmer, 1993). The mutant was found to produce detectable levels of pectate lyase activity which resulted from a second set of plant-inducible enzymes synthesized only *in planta*. Similar genetic evidence for fungal pathogens is scarce. However, the targeted disruption of the single endopolygalacturonase gene of *Cochliobolus carbonum* suggested that the enzyme is not required for pathogenicity on maize (Scott-Craig *et al.*, 1990).

Certain bacterial plant pathogens, *Pseudomonas*, *Xanthomonas* and some *Erwinia* spp., contain large clusters of plant-regulated genes termed *h*ypersensitive, *r*esponse and *p*athogenicity (*hrp*) genes which are indispensable for growth and pathogenicity in plants (Willis *et al.*, 1991). The *hrp* genes are also required for eliciting the HR in host plants in incompatible interactions or in non-host plants (Atkinson, 1993). Introduction of the *Erwinia amylovora hrp* gene cluster into *Escherichia coli* enabled this non-pathogen to elicit an HR in tobacco, by producing a 44 kDa proteinaceous elicitor, named harpin, the product of the *hrpN* gene (Wei *et al.*, 1992). The functions of the *hrp* genes are still poorly understood. Some share homology to virulence determinants of animal-pathogenic bacteria and thus, in analogy, may be part of the secretion machinery involved in the export of eliciting molecules (Gough *et al.*, 1992; Salmond, 1994).

Although the initial entry into the host is an essential requirement for a pathogen to be successful, subsequent to this, pathogens have to escape or to cope with the activated plant defense responses in order to ultimately colonize their hosts. Strategies employed by pathogens include weakening of the host by toxins and/or inhibiting or inactivating parts of the hosts' defense machinery. The best characterized example of this kind is the detoxification and inactivation of the pea phytoalexin, pisatin, by a pisatin demethylase of *Nectria haematococca* (VanEtten *et al.*, 1989; Schäfer, 1994). When the pisatin demethylase gene was introduced into *Cochliobolus heterostrophus*, a fungal pathogen of maize but not of pea, it enabled *Cochliobolus heterostrophus* to successfully attack pea plants and thus to be virulent on this new host (Schäfer *et al.*, 1989; Oeser and Yoder, 1994).

Toxins produced by bacterial and fungal pathogens have also been implicated in virulence, some of which exhibit a high degree of specificity, allowing pathogenesis only on certain plant cultivars or species (Keen, 1992). The best understood mechanism of host-selective toxicity is that of T toxin from *Cochliobolus heterostrophus* (anamorph *Helminthosporium maydis*) race T which binds to a protein of the inner mitochondrial membrane of the host (Levings and Siedow, 1992). Conversely, the ability of a plant to tolerate or detoxify a toxin may determine resistance (see below). Suppressors, in contrast to toxins, do not cause disease symptoms and their presence has been inferred from the absence of host defense responses in infected tissue or by their suppression *in vitro* in model systems (Knogge, 1991; Collinge *et al.*, 1994). However, despite physiological and biochemical evidence for their existence, the structures and modes of action of putative suppressors are unknown and their role in plant–pathogen interactions remains to be genetically proven.

B. AVIRULENCE FUNCTIONS OF PATHOGENS AND PLANT RESISTANCE GENES

In some cases, pathogen-encoded virulence functions can also play a key role in activating plant defense responses. Cell wall hydrolases, for example, are not only important for the pathogen to breach the plant cell wall thereby gaining access to vital nutrients, but are also essential components of the recognition process by which the plant senses the presence of a pathogen. This recognition is a prerequisite for the activation of the plant's own defense program. In most cases it appears that plants do not recognize the extracellular cell wall hydrolases directly, with the exception of xylanase (Sharon *et al.*, 1993), but rather respond to the enzymatic products they release. This has most convincingly been demonstrated for the oligogalacturonides (Ryan and Farmer, 1991). The efficiency of these plant-derived endogenous signals to initiate the defense program can be potentiated when combined with pathogen-derived, exogenous signals, commonly referred to as elicitors. The exogenous elicitors of plant defense responses differ widely in their chemical nature, ranging from proteins, oligosaccharides, glycoproteins to lipids and derivatives thereof (Ebel and Cosio, 1994). Most of the pathogen-derived elicitors identified to date are non-specific, that is, they induce various defense responses in a large variety of plant cultivars and species.

In contrast to these, certain bacterial and fungal avirulence (*avr*) genes have been shown to encode race-specific elicitors which efficiently trigger defense responses only in plant cultivars harboring the complementary resistance (*R*) genes (Keen, 1992, Kamoun *et al.*, 1993; Marmeisse *et al.*, 1993; Pryor and Ellis, 1993; Knogge *et al.*, 1994). One can therefore envision that cultivar-race-specific recognition may occur via the direct interaction of the *avr* gene product with the corresponding plant *R* gene product. However, only in the case of the *avr9* gene from *Cladosporium fulvum* could the processed protein product be shown to directly induce an HR in a cultivar-specific manner (Marmeisse *et al.*, 1993). Most of the numerous bacterial *avr* genes identified and cloned over the past 10 years probably encode cytoplasmic proteins, and hence are unlikely to act directly as host recognition determinants (Keen, 1992; Salmond, 1994). Some may have enzymatic activities by which pathogen metabolites are converted into elicitors as has been suggested for the *avr*D gene from *Pseudomonas syringae* (Keen, 1992), whereas others may play a role in elicitor transport or secretion, e.g. the *hrp* gene products.

Even less is known on how major plant *R* genes determine plant resistance. Long and intensive efforts to isolate *R* genes have only recently proven to be successful. To date, six plant *R* genes have been cloned (see De Wit, this volume), the maize *Hml* locus conferring resistance to *Cochliobolus carbonum* (Johal and Briggs, 1992), the tomato *Pto* locus and the *Arabidopsis* *RPS2* gene conferring resistance to *Pseudomonas syringae* pv. *tomato* (Martin *et al.*, 1993; Bent *et al.*, 1994; Mindrinos *et al.*, 1994), the tobacco gene *N* mediating resistance to tobacco mosaic virus (Whiteham *et al.*, 1994), the flax L^6 gene conferring resistance to flax rust (Lawrence *et al.*, 1994), and the tomato *Cf9* gene conferring resistance to *Cladosporium fulvum* (Jones *et al.*, 1994). The *Hml* gene appears to encode a reductase which detoxifies the HC toxin of *Cochliobolus carbonum* thereby rendering the plant resistant (Meeley *et al.*, 1992). The role of the other *R* gene products remains to be elucidated. Preliminary data, derived purely from amino acid sequence comparisons, suggest that they may be components of the signaling pathways leading to the activation of plant defense responses.

III. CLASSES OF PLANT DEFENSE RESPONSES

Over the past decades, considerable advances in our understanding of natural mechanisms associated with plant resistance to their naturally occurring aggressors have been made. From the analysis of many plant–pathogen interactions, a huge body of evidence has accumulated on different defense mechanisms (Fig. 1). For some, a function in inhibiting pathogen development has been clearly demonstrated (e.g. phytoalexins, hydrolytic enzymes, proteinase inhibitors), whereas for others the evidence remains circumstantial (e.g. cell wall modification, HR). Most of the inducible defense responses are the result of transcriptional activation of specific genes, the regulation of which has been studied extensively in both intact plant–pathogen interactions and in model systems consisting of cultured plant cells and elicitors. Rather than describing in detail the many individual defense reactions that have been identified in the various systems, we will collectively discuss general aspects of three classes of defense response: (1) immediate, early plant defense responses, considered to be involved in recognition and initial signaling events along with constituting the first line of defense towards pathogen ingress; (2) locally activated defense mechanisms, which are thought to have direct adverse effects on the pathogens; and (3) systemically induced defense responses and their relationship to systemic acquired resistance (SAR). This classification is based on the tem-

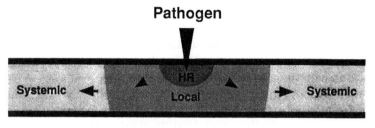

Pathogen

Fig. 2. Schematic representation of three major classes of plant defense response observed after fungal infection and their relative timing of induction. See text for details.

poral and spatial pattern of expression of different defense responses as observed in several systems (Fig. 2). However, since certain responses have dual functions or may exhibit variable expression patterns in different systems or under different conditions, this classification is somewhat arbitrary. Since, in fact, very few model systems have been studied in sufficient detail with respect to the expression of the many different defense responses, no conclusions on causal links or successive signaling mechanisms should be inferred.

A. IMMEDIATE, EARLY DEFENSE RESPONSES: RECOGNITION AND INTRACELLULAR SIGNALING

It is obvious that plant cells must have both the capability to recognize a potential pathogen, and the ability to transduce this information, within the cell and to neighboring cells, in order to mount an efficient defense response. Our knowledge of the mechanisms by which this is achieved remains still rather fragmentary. Nevertheless, intensive studies within the last few years are starting to reveal important components of the plant signal perception and signal transduction pathways. An important step toward this goal was the isolation, purification and structural elucidation of homogeneous elicitor-active components from various fungal and bacterial sources. Examples are the heptaglucoside (β-glucan) elicitor from *Phytophthora megasperma*, the 42 kDa glycoprotein from the same fungus and the oligopeptide of 13 amino acids derived therefrom, the oligo-1,4-α-galacturonides, the necrosis-inducing peptides from *Cladosporium fulvum* and *Rhynchosporium secalis*, and the protein harpin from *Erwinia amylovora* (for review see Ebel and Cosio, 1994). High-affinity binding sites for several of these pathogen-derived carbohydrate and protein elicitors have been identified on the plasma membranes of soybean, tomato, parsley and rice cells (Basse *et al.*, 1993; Shibuya *et al.*, 1993; Ebel and Cosio, 1994; Nürnberger *et al.*, 1994). Hopefully, isolation and characterization of such plant cell receptors will be forthcoming within the next few years.

Based on studies with cultured parsley cells and the proteinaceous elicitor derived from the soybean pathogenic fungus *Phytophthora megasperma*, a model of the intracellular signal transduction pathway leading to the initiation of defense responses upon ligand binding to the plant cell receptor has been proposed (Fig. 3). The most rapid responses of parsley cells to elicitor treatment detected so far are changes in the H^+, K^+, Cl^- and Ca^{2+} fluxes across the plasma membrane and the formation of H_2O_2 (oxidative burst), which occur within 2–5 min (Nürnberger *et al.*, 1994). After 5–30 min, Ca^{2+}-dependent phosphorylation of proteins *in vivo* was observed along with the rapid biosynthesis of ethylene and the transcriptional activation of the most rapidly induced defense-related genes (Somssich *et al.*, 1989; Dietrich *et al.*, 1990; Boller, 1991). On a temporal basis, the oxidative burst and protein phosphorylation/dephosphorylation appear to be downstream of ion channel activation, but their possible causal relationship with other putative components of the signal transduction chain, i.e. GTP-binding proteins, inositol phosphates, cAMP, etc., leading to the activation of defense responses remain to be elucidated. Subsequent to these initial events, activation of other defense genes and accumulation of phytoalexins is detected.

Ion fluxes, protein phosphorylation and synthesis of active oxygen species (oxidative burst) have also been observed in many other plant species after infection or elicitor treatment (reviewed in Atkinson, 1993; Dixon *et al.*, 1994; Ebel and Cosio, 1994; Mehdy, 1994). Their proposed function as cellular signaling components, however, is based only on correlative evidence obtained from studies with model systems, such as cultured cells. The obvious advantage of such systems is that they allow a more precise temporal analysis of induction processes, the application of specific inhibitors of signaling compon-

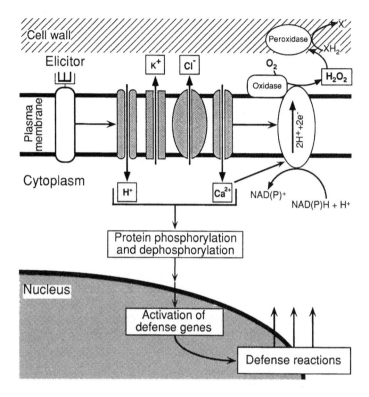

Fig. 3. Model of components involved in signal tranduction pathways leading to plant defense gene activation. Adapted from Nürnberger *et al.* (1994)

ents (e.g. ion channel blockers, ionophores, protein kinase and phosphatase inhibitors) and other technical manipulations to which infected intact plant tissues are not easily amenable. This is particularly true when very rapid responses, which occur within minutes or even seconds after stimulation of cells with elicitor, are being investigated for their participation in the signaling cascade (Apostol *et al.*, 1989; Baker *et al.*, 1993; Felix *et al.*, 1993; Legendre *et al.*, 1993; Viard *et al.*, 1994). However, not in all cases was the initial signaling event tightly correlated with the subsequent induction of a defense response, such as phytoalexin synthesis or defense gene activation, indicating that alternative signaling pathways may exist (Devlin and Gustine, 1992; Felix *et al.*, 1993).

Although many of the very rapid biochemical responses detected in plant cells upon elicitor treatment most likely represent components of the induced signaling pathways, some may also participate directly in plant defense. Because of the cytotoxicity of active oxygen species, such as the superoxide anion (O_2^-) and hydrogen peroxide (H_2O_2), and the rapidity of their production, the oxidative burst has been suggested to be the *first line of defense* against pathogen invasion by directly killing the pathogen and/or slowing down its ingress by stiffening the plant cell wall via oxidative cross-linking of specific structural proteins (Apostol *et al.*, 1989; Bradley *et al.*, 1992; Mehdy, 1994). In accordance with such a function is the observation that active oxygen species are generated in numerous plant tissues in response to pathogen infection (Tzeng and DeVay, 1993).

The toxicity of active oxygen species may also contribute to host cell death and hence they have been implicated in playing a role in the rapid collapse and death of infected cells that occurs during HR (Tzeng and DeVay, 1993; Mehdy, 1994). This has recently been substantiated by results demonstrating that H_2O_2 may in fact be a trigger of hypersensitive cell death (Levine *et al.*, 1994). The HR appears to be a multicomponent, genetically determined process that is still poorly understood at the molecular level. It is observed in many incompatible cultivar–race–specific interactions as well as in infected non-host plants and it is believed to restrict the pathogen by depriving it of the necessary nutrients required for growth and development. The HR appears to involve irreversible membrane damage, the result of early electrolyte leakage and lipid peroxidation (Atkinson, 1993; Collinge *et al.*, 1994). Cytological studies using video microscopy techniques for live examination of the dynamic aspects of early events in plant–fungus interactions revealed that cytoplasmic rearrangements precede the HR, wall apposition (papillae formation) and other activated defense responses (Freytag *et al.*, 1994). Since many of the general physiological characteristics of HR are common to various plant species, it is reasonable to assume that the biochemical and molecular basis of hypersensitivity is similar in the different pathosystems studied.

B. LOCAL DEFENSE RESPONSES

Closely associated with the HR are diverse biochemical plant responses, many of which involve the *de novo* synthesis of proteins as a consequence of direct pathogen-mediated transcriptional activation of the corresponding genes (Bowles, 1990; Dixon and Harrison, 1990; Cutt and Klessig, 1992; Collinge *et al.*, 1994; Somssich, 1994; Kombrink and Somssich, 1995). By virtue of their catalytic activities, these proteins either directly or indirectly restrict growth and development of invading pathogens. They include key enzymes of general phenylpropanoid metabolism, such as phenylalanine ammonia-lyase (PAL) and 4-coumarate:CoA ligase (4CL), the innumerable proteins involved in the biosynthesis of phytoalexins and other secondary metabolites, hydroxyproline-rich glycoproteins (HRGP) and glycine-rich proteins (GRP), intra-and extra-cellular PR proteins, including chitinases and 1,3-ß-glucanases, peroxidases, lipoxygenases, proteinase and polygalacturonase inhibitors, and antimicrobial proteins. In addition, numerous cDNAs and genes have been isolated by differential screening methods to which no biochemical function can yet be attributed but which are nevertheless strongly induced upon infection (Somssich *et al.*, 1989; Taylor *et al.*, 1990; Pontier *et al.*, 1994).

Figure 4 summarizes components of various biochemical pathways that have been shown to be transcriptionally activated in parsley cells by an elicitor of *Phytophthora megasperma* and in parsley leaves upon inoculation with spores of the same fungus (Schmelzer *et al.*, 1989; Somssich *et al.*, 1989; Kawalleck *et al.*, 1992, 1993, 1995). Figure 4 also emphasizes the important role of the phenylpropanoid metabolism in plant defense, which has been found to be activated in virtually every plant–pathogen system analyzed to date (Hahlbrock and Scheel, 1989; Nicholson and Hammerschmidt, 1992). From its various branch pathways a large variety of compounds with diverse functions are derived, including pigments, antibiotics (phytoalexins), UV protectants, signal molecules and structural components, such as lignin, suberin, and other cell wall constituents. Incorporation of phenolic compounds into the cell wall is one of the most rapidly occurring defense reactions. It is readily detectable in infected tissue, due to the strong autofluorescence of phenolics under blue or UV light (Jahnen and Hahlbrock, 1988; Schmelzer *et al.*, 1989; Schröder *et al.*, 1992; Freytag *et al.*, 1994). These and other modifications of the cell wall, such as lignification or incorporation of structural proteins (e.g. HRGP), are presumably catalyzed by peroxidases and render the cell wall less permeable to pathogens. Phytoalexins, a diverse group of chemical compounds with broad spectrum antibiotic activities, are synthesized by plants in response to infection and have been studied extensively over the past decade (Dixon and Harrison, 1990; Nicholson and Hammerschmidt, 1992). Each plant species produces a characteristic set of phytoalexins which are derived from secondary metabolism and most frequently constitute phenylpropanoid, terpenoid, or polyacetylene derivatives. Their biosynthetic enzymes are usually induced sequentially, as demonstrated in many cases, including PAL, 4CL and S-adenosyl-L-methionine:bergaptol O-methyltransferase (BMT) which are involved in the formation of furanocoumarin phytoalexins in parsley (Fig. 4) (Hahlbrock and Scheel, 1989). One rapid response, the synthesis of callose, is not regulated at the transcriptional level but via allosteric activation of the biosynthetic enzyme, $1,3$-β-glucan synthase (Kauss, 1987). Callose deposition on cell walls in the form of papillae is a highly localized defense mechanism occurring around fungal penetration sites (Gross *et al.*, 1993).

Since HR mediated resistance initiates from a clearly defined cellular position within the plant tissue, both the temporal and spatial expression patterns of defense mechanisms are bound to be important. Hence, their rapid and localized activation at the site of ingress following initial recognition of the pathogen, can be considered a *second line of defense* that is essential to limit invaders to relatively small tissue areas. Detection of such expression patterns in infected tissue requires sensitive and specific histological methods. Since

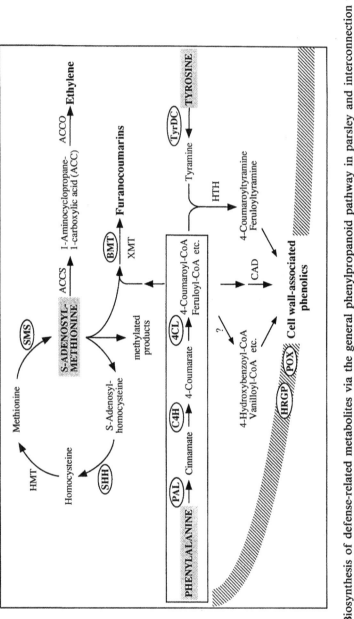

Fig. 4. Biosynthesis of defense-related metabolites via the general phenylpropanoid pathway in parsley and interconnection with other selected elicitor-inducible reactions. Encircled are identified enzymes for which genes have been isolated and for which transcriptional control has been demonstrated. Abbreviations: 4CL, 4-coumarate:CoA ligase; ACCO, ACC oxidase (ethylene-forming enzyme); ACCS, ACC synthase; BMT, *S*-adenosyl-L-methionine:bergaptol *O*-methyltransferase; C4H, cinnamate 4-hydroxylase; CAD, cinnamyl-alcohol dehydrogenase; HMT, homocysteine methyltransferase; HRGP, hydroxyproline-rich glycoproteins; HTH, hydroxycinnamoyl-CoA:tyramine hydroxycinnamoyltransferase; PAL, phenylalanine ammonia-lyase; POX, peroxidase; SHH, *S*-adenosyl-L-homocysteine hydrolase; SMS, *S*-adenosyl-L-methionine synthetase; TyrDC, tyrosine decarboxylase; XMT, *S*-adenosyl-L-methionine:xanthotoxol *O*-methyltransferase.

most phytoalexins are not visible, it has been difficult to determine their precise localization within tissues. However, in special cases and with specific techniques it has been possible to demonstrate their local accumulation at sites of fungal ingress (Hahn *et al.*, 1985; Jahnen and Hahlbrock, 1988; Snyder and Nicholson, 1990). *In situ* localization studies of proteins and mRNAs have clearly revealed that plant cells adjacent to those that are in direct contact with the pathogen, and thus undergo hypersensitive cell death, respond massively by activation of numerous defense mechanisms. One can discriminate between responses which occur within a rather limited number of cells in the vicinity of infection sites (local responses), and others which are detectable also at a distance, within all cells of a given tissue or throughout the entire plant (systemic responses).

Activated defense-related genes that appear locally restricted include those encoding PAL, 4CL, BMT, tyrosine decarboxylase (TyrDC), the intracellular PR proteins (*PcPR1* from parsley and *PRP1* from potato), and an anionic peroxidase, to name just a few (Schmelzer *et al.*, 1989; Taylor *et al.*, 1990; Schröder *et al.*, 1992; Kombrink *et al.*, 1993; Freytag *et al.*, 1994; Kawalleck *et al.*, 1995). Such *in situ* localization studies also demonstrated that, in the case of the local defense response, the induced levels of detectable gene expression are the same for all cells over the whole affected area and that there is a rather sharp demarcation between fully activated and completely unaffected cells (Schmelzer *et al.*, 1989). This suggests that there is a threshold phenomenon in the signaling mechanism in response to a pathogen. Most likely, all responding cells within the affected area will ultimately undergo cell death. The expression pattern observed after wounding the tissue is clearly distinct. Although many defense-related genes are also wound-inducible, the signals generated result in high level expression of these genes only in cells adjacent to the wound site, whereas expression levels continue to decline gradually the more distant the cells are from the injured area (Schmelzer *et al.*, 1989). With respect to the enzymes of phytoalexin biosynthesis, their local induction does not indicate whether phytoalexins themselves accumulate in the first infected cell.

Despite a strong correlation of HR development with the rapid and local activation of plant defense-related genes, there are very few data which demonstrate that there is indeed a causal relationship between the two. Most of the defense-related genes can be activated by elicitors that do not cause an HR or are expressed to similar levels in compatible and in incompatible interactions despite the occurrence of an HR only in the incompatible case (Schröder *et al.*, 1992; Atkinson, 1993; Jakobek and Lindgren, 1993; van Gijsegem *et al.*, 1995, in press). In fact, to date there exists only one report of an induced plant gene, *hsr203J* from tobacco, encoding a 37.5 kDa protein of unknown function, the expression of which is absolutely dependent on the HR (Pontier *et al.*, 1994). Additionally, it is extremely difficult to demonstrate whether expression of defense-related genes actually occurs within those few cells undergoing hypersensitive cell death, or whether gene activation is restricted only to the neighboring cells. Recently, expression of the *PcPRl* gene was demonstrated to occur in a fungus-infected parsley cell by *in situ* RNA hybridization techniques (Gross *et al.*, 1993). However, it was technically impossible to determine the eventual fate of this cell.

An extremely promising approach to elucidate the molecular basis of the HR is the employment of mutants showing aberrant regulation of the HR. Several such mutants have been observed in different plant species, displaying visible phenotypes that resemble the lesions caused by pathogen attack (Wolter *et al.*, 1993; Godiard *et al.*, 1994). Genes for a number of such mutants have recently been identified in *Arabidopsis* which correlate with the expression of histochemical and molecular markers of plant defense responses (Dietrich *et al.*, 1994; Greenberg *et al.*, 1994). The obvious question to be resolved is whether the loci responsible for the mutant aberrant cell death phenotypes are also involved in the normal process of disease resistance.

Finally, despite overwhelming evidence that HR is a highly important component of the defense machinery of the plant, in a number of instances HR-independent resistance is observed (Atkinson, 1993; Knogge *et al.*, 1994). Thus, plants obviously possess additional means to deter pathogen invasion.

C. SYSTEMIC DEFENSE RESPONSES

In many plant species, local infection by a variety of pathogens leads to the induction of general disease resistance in uninfected parts of the plant (Madamanchi and Kuc, 1991; Hammerschmidt, 1993; Kessmann *et al.*, 1994). Typically, this resistance is effective against a broad range of pathogens, including viruses, bacteria and fungi, and can last for several weeks or even months. This phenomenon, termed induced resistance or systemic acquired resistance (SAR), has been known for many years but it is only recently that it has been correlated with the systemic activation of several specific plant defense responses via a signal transduction pathway initiated locally at the initial site of pathogen attack (Kessmann *et al.*, 1994). However, only relatively few of the numerous pathogen-induced defense mechanisms are strictly correlated with SAR. They can be considered an additional component of the plant's integrated disease resistance repertoire, the *third line of defense* which is directed against subsequent invaders.

SAR seems to be a general property of plants and its underlying biochemical and molecular mechanisms have been most extensively studied in tobacco (*Nicotiana tabacum*), cucumber (*Cucumis sativus*) and *Arabidopsis* (Madamanchi and Kuc, 1991; Uknes *et al.*, 1993; Cameron *et al.*, 1994; Kessmann *et al.*, 1994). In tobacco it has been clearly demonstrated that the onset of SAR correlates with the induction of extracellular PR proteins, more specifically, at least nine families of genes are coordinately and systemically activated in response to local infection with tobacco mosaic virus (Ward *et al.*, 1991). Proteins encoded by these genes include 1,3-β-glucanases, chitinases, chitin-binding proteins, thaumatin-like proteins, and PR-1, all of which have anti-fungal activity (Kombrink and Somssich, 1995). The same set of genes was coordinately activated by application of salicylic acid (SA), a putative endogenous translocated signal (see below), or by 2,6-dichloroisonicotinic acid, a synthetic regulator of induced resistance (Kessmann *et al.*, 1994). Interestingly, however, induction was restricted to the acidic isoforms of the PR proteins (Brederode *et al.*, 1991; Ward *et al.*, 1991). This would suggest that the basic PR proteins are not involved in SAR although only these isoforms have been shown to possess marked antifungal activity (Sela-Buurlage *et al.*, 1993). In contrast, systemic induction of 1,3-β-glucanases and chitinases in potato was clearly demonstrated to include both acidic and basic isoforms as revealed by localized tissue sampling, immunohistochemistry, and *in situ* mRNA localization (Schröder *et al.*, 1992; R. Büchter, M. Müller, E. Kombrink, unpublished results). SAR in cucumber induced by viruses, bacteria, fungi, or chemicals, has been correlated with a large variety of biochemical responses partly distinct from those observed in tobacco. In addition to a unique class III chitinase, increased levels of lignification, peroxidase, hydroxyproline-rich glycoproteins, lipoxygenase and SA have been reported (Madamanchi and Kuc, 1991; Avdiushko *et al.*, 1993; Hammerschmidt, 1993; Lawton *et al.*, 1994a). Likewise, in *Arabidopsis* SAR is correlated with the activation of a subset of genes identified in tobacco, excluding chitinases (Uknes *et al.*, 1992, 1993). However, despite their apparent association, the demonstration of a causal relationship between SAR and the induced expression of these genes remains outstanding. *Arabidopsis* should prove an extremely useful system for this, allowing the genetic dissection of the disease resistance mechanisms associated with SAR.

Probably the best studied example of systemic gene expression relates to the induction of proteinase inhibitors I and II (PI-I and II) in tomato and potato by localized insect damage or mechanical injury (Ryan, 1990, 1992). The inhibitors are potent inactivators of proteolytic enzymes found in the digestive tract of animals and have been demonstrated to protect plants against foreign invaders, particularly insects. Induction of PIs by microbial pathogens has also been demonstrated in melon, tomato and tobacco (Roby *et al.*, 1987; Geoffroy *et al.*, 1990; Pautot *et al.*, 1991; Linthorst *et al.*, 1993), although functional data as to their possible role in disease resistance are still missing. Induction of SAR by pathogen preinoculation on the one hand, and insect resistance by wounding on the other hand, seem to be mutually exclusive processes indicating that different signaling pathways are activated by the two stimuli, acting via SA and systemin, respectively (see below).

IV. REGULATION OF DEFENSE GENE EXPRESSION

A. SIGNALS INVOLVED IN DEFENSE GENE ACTIVATION

The nature of the signals participating in both the local and systemic activation of defense mechanisms remains to be unequivocally determined. In recent years, numerous putative signal molecules have been identified which should greatly facilitate our understanding of the signal generation mechanisms and transduction pathways employed in disease and pest resistance (Enyedi *et al.*, 1992; Hammerschmidt, 1993).

Ethylene has long been known to be synthesized by plants under stress conditions, including pathogen infection and mechanical wounding (Boller, 1991). Ethylene biosynthesis proceeds from *S*-adenosyl-L-methionine (SAM), an intermediate of the activated methyl cycle, via 1-aminocyclopropane-1-carboxylic acid (ACC) which is formed by the inducible enzyme ACC synthase (Fig. 4). A signaling function of ethylene has been imposed due to its high mobility (as a gas it can readily diffuse from its site of synthesis) and its capacity to induce numerous defense responses, particularly hydrolytic enzymes (Boller, 1991). However, recent experiments using norbornadien, an inhibitor of ethylene action, and ethylene-response mutants of *Arabidopsis* indicate that ethylene is not directly involved in signal transduction leading to PR protein accumulation and SAR induction, nor does it cause accumulation of salicylic acid (Silverman *et al.*, 1993; Lawton *et al.*, 1994b). Although ethylene influences diverse physiological plant processes and can induce some of the defense-related genes associated with local defense and SAR, convincing evidence of its role in the signal transduction pathway leading to plant defense is lacking.

Some exciting new insights have been gained from research centered around the signal molecule salicylic acid (SA). SA appears to be involved in numerous plant processes, such as flowering, thermogenesis, as well as disease resistance (Enyedi *et al.*, 1992; Malamy and Klessig, 1992). As demonstrated for tobacco, cucumber and *Arabidopsis*, infection with viruses, bacteria or fungi results in an increase of SA levels suggesting that it is an endogenous signal in the defense response (Enyedi *et al.*, 1992; Malamy and Klessig, 1992; Hammerschmidt, 1993; Uknes *et al.*, 1993). This is supported by the fact that exogenous application of salicylate induces many of the biochemical responses associated with SAR (Ward *et al.*, 1991; Enyedi *et al.*, 1992; Uknes *et al.*, 1992) and by the observation that the high degree of resistance of the amphidiploid hybrid *Nicotiana glutinosa* × *Nicotiana debneyi* towards various types of pathogens correlates with elevated levels of endogenous SA and high constitutive levels of PR proteins (Yalpani *et al.*, 1993). Even more convincing evidence comes from transgenic tobacco plants harboring the bacterial *nah*G gene which encodes salicylate hydroxylase, an enzyme converting SA into catechol, an inactive compound. These plants did not accumulate SA and were defective in their ability to manifest SAR (Gaffney et al., 1993). Controversy still exists whether SA itself is the long-distance, mobile signal. Removal of the inducer leaf prior to a significant increase in SA levels did not prevent SAR and suggested the existence of other systemic signals (Hammerschmidt, 1993). The same conclusion was derived from grafting experiments performed between wild-type tobacco and transgenic plants expressing the *nah*G gene. Although unable to accumulate SA, transgenic rootstocks were fully capable of inducing SAR in non-transgenic scions, indicating that SA is not the translocated molecule (Vernooij et al., 1994). In contrast, recent $^{18}O_2$-labeling experiments provide good evidence that SA is indeed systemically translocated in tobacco. Some 60% of the SA accumulating in untreated upper leaves was found to originate from lower, inoculated leaves (V. Shulaev, I. Raskin, personal communication). Interestingly, UV light and ozone also stimulate accumulation of SA, PR proteins and resistance in tobacco whereas wounding does not (Brederode et al., 1991; Enyedi et al., 1992; Yalpani et al., 1994).

Klessig and co-workers have recently purified and cloned an SA-binding protein and found it to exhibit catalase activity (Chen et al., 1993a, b). Catalase effectively inactivates H_2O_2 by converting it to H_2O and O_2. SA inhibited *in vitro* catalase activity and induced elevated levels of H_2O_2 in tobacco leaves. The conclusion drawn from these and other experiments is that the action of salicylate in SAR is mediated by elevated levels of H_2O_2. Thus, active oxygen species may play a pivotal role in both the rapid, local and the systemic defense mechanisms leading to plant resistance.

Two other endogenous plant compounds, jasmonic acid (JA) and its volatile methyl ester, methyl jasmonate (MeJA), have also been shown to activate systemic plant defense responses and thus to be potentially important in signaling (Enyedi et al., 1992; Hammerschmidt, 1993). It has been shown that MeJA is a potent inducer of proteinase inhibitors (PI) in tomato, tobacco, and alfalfa leaves, being even more effective than wounding (Farmer et al., 1992; Ryan, 1992). Furthermore, evidence was presented that precursors of jasmonic acid biosynthesis, including free linolenic acid, can act as signals for PI accumulation whereas structurally related compounds that are not precursors were inactive (Farmer and Ryan, 1992). So far, few reports exist on the possible involvement of JA and MeJA in signaling of defense responses directed against pathogens (Gundlach et al., 1992; Mueller et al., 1993; Xu et al., 1994). Biosynthesis of JA and MeJA involves lipoxygenase-catalyzed peroxidation of fatty acids which are thought to be released from damaged membranes. The frequently observed increases in lipoxygenase gene expression and enzyme activities after wounding, elicitor treatment or pathogen infection lends additional support for a role of lipid-based signals in systemic gene activation although this response is relatively slow compared to activation of other defense-related genes (Atkinson, 1993; Meier et al., 1993; Melan et al., 1993; Fournier et al., 1994; Geerts et al., 1994).

A search for the PI-inducing factor ultimately led to the isolation of a small, 18-amino-acid peptide, named systemin, from tomato tissue. Systemin is the first peptide hormone-like molecule identified so far in plants (Ryan, 1992). Apart from inducing PI gene expression, it has been shown that systemin is readily transported through the phloem from wounded leaves to upper, unwounded leaves in tomato plants (Narváez-Vásquez et al., 1995). Furthermore, constitutive expression of antisense systemin mRNA in transgenic plants, abolished wound induction of PIs in leaves and resulted in a decrease in resistance of these plants towards larvae of the Lepidoptera predator, *Manduca sexta* (Orozco-Cardenas et al., 1993). Conversely, expression of a prosystemin cDNA in tomato plants resulted in constitutive accumulation of extraordinarily high PI levels without wounding or chemical induction, also in untransformed (wild-type) scions grafted onto a transgenic rootstock (McGurl et al., 1994). These results not only strongly support the proposed role of systemin as the mobile wound signal, they also demonstrate that external stimuli and/or other signaling mechanisms, such as the proposed electrical activity (Wildon et al., 1992), are not necessary for induction of PI expression.

Oligogalacturonides were the first plant-derived signal molecules that were correlated with systemic PI induction, and they are also known as endogenous elicitors of many other plant defense responses (Ryan and Farmer, 1991; Ebel and Cosio, 1994). However, their limited mobility within plants restricts their function as systemic signals (Ryan and Farmer, 1991). They are released from plant cell walls by pectinolytic enzymes of pathogens. It has been proposed that degradation of elicitor-active molecules (degree of polymerization 10–15) is prevented by plant-encoded polygalacturonase-inhibiting proteins (PGIP), thus allowing the plant to recognize an invader and to efficiently mount its defense (Bergmann et al., 1994). The local induction of PGIP around infection sites supports its proposed function in generating signals for local activation of other defense responses (Bergmann et al., 1994). However, this does not preclude the possibility that PGIP also directly retards pathogen ingress by inactivating essential virulence functions. Interestingly, PGIP shows a certain degree of sequence similarity to the *Cf9* resistance gene product of tomato (Jones et al., 1994). The participation of other plant derived proteins, namely 1,3-β-glucanases and chitinases, in the release of elicitors from pathogens has recently been reviewed elsewhere (Ebel and Cosio, 1994; Kombrink and Somssich, 1995).

B. REGULATORY COMPONENTS OF DEFENSE GENES

In nearly all systems studied, transcriptional activation of defense genes is essential for the plant to mount an efficient defense response towards pathogen attack (Dixon and Harrison, 1990; Cutt and Klessig, 1992; Somssich, 1994). Interest in understanding, in detail, the molecular mechanisms underlying pathogen-induced plant gene activation is not purely academic. Rather, such knowledge is a prerequisite for the development of safe and sound strategies to genetically engineer resistance in important crop plants which otherwise are incapable of preventing pathogen ingress. Since many plant defense-related genes are also regulated in a cell-type-, organ- and development-specific manner, dissection of the regulatory regions of such genes is required in order to identify DNA elements important solely for pathogen inducibility. Use of such clearly defined promoter sequences would be ideal to drive the expression of genes encoding antimicrobial compounds at the site of pathogen invasion.

Progress in this field of research has moved rather slowly. Important regulatory regions within the promoters of defense-related genes have been identified; however, in only a limited number of cases have functionally defined *cis*-acting DNA elements been actually pinpointed (recently reviewed in Dixon and Harrison (1990), Dixon *et al.* (1994) and Somssich (1994)). Even less is known about the protein factors which specifically interact with these DNA motifs. In parsley, two cDNAs, one coding for a protein factor which binds to the PAL promoter and another encoding a homeodomain protein, which specifically interacts with the PR2 promoter, have recently been isolated (da Costa e Silva *et al.*, 1993; Korfhage *et al.*, 1994). Additionally, two proteins, KAP1 and KAP2, which specifically interact with an element within the bean chalcone synthese promoter have been purified to apparent homogeneity (Dixon *et al.*, 1994). Such studies are often hampered by the presence of functionally redundant DNA elements in plant promoters and by the lack of an accurate *in vitro* transcription system for assessing the requirement of identified DNA binding proteins in efficient and specific transcription of defense-related plant genes.

V. DEFENSE RESPONSES AND RESISTANCE

A. INHIBITION OF PATHOGENS *IN VITRO*

The classification of a biochemical plant component as being involved in defense against invading microorganisms is usually based on three criteria: (1) its antimicrobial activity which is usually demonstrated by *in vitro* growth inhibition of test microbes; (2) its induction or accumulation in response to pathogen challenge; and (3) its differential expression behavior in cultivar–race–specific interactions. The classical example for such components are the phytoalexins. Their mechanisms of action are presumably as diverse as their structures. Likewise, *in vitro* growth inhibition of fungal and bacterial pathogens has been demonstrated for a large variety of plant proteins, including chitinases, 1,3-β-glucanases, permatins (thaumatin-like proteins, osmotin), thionins, lectins, ribosome-inactivating proteins, and other proteinaceous inhibitors and toxin-like peptides (reviewed in Kombrink and Somssich, 1995).

Most of these proteins are induced in vegetative plant organs by infection but some are also abundant seed constituents. Chitinases and 1,3-β-glucanases exert their inhibitory function by hydrolyzing structural components of fungal cell walls, chitin and 1,3-β-glucans, as supported by morphological studies and the release of soluble carbohydrates, but the inhibitory capacity varies considerably between different isoforms of both enzymes (Mauch *et al.*, 1988; Arlorio *et al.*, 1992; Sela-Buurlage *et al.*, 1993; Kombrink and Somssich, 1995). Permatins and thionins apparently increase plasma membrane permeability of fungi, whereas ribosome-inactivating proteins catalytically inactivate non-self ribosomes (Stirpe *et al.*, 1992; Vigers *et al.*, 1992; Bohlmann, 1994). However, for many putative defense responses the physiological role in inhibiting pathogen growth or development has not been critically tested, whereas for others, *in vitro* experiments are technically not feasible. This is the case, for example, for topically expressed or complex responses, such as papilla formation, cross-linking and reinforcement of cell walls, or the HR. Finally, it has been demonstrated that combinations of different pathogen inhibitors can act synergistically, which in view of their different modes of action may not be surprising.

B.   EXPRESSION PATTERNS IN INFECTED PLANTS

To elucidate the mechanisms that determine disease resistance, numerous plant–pathogen interactions showing cultivar–race–specific resistance have been analyzed for the expression of putative defense responses. In many cases it has been shown that various defense-related transcripts, proteins, or products (i.e. phytoalexins) accumulate more rapidly and/or to higher levels in incompatible interactions as compared to compatible ones (Dixon and Harrison, 1990; Kombrink and Somssich, 1995). In various instances it has even been demonstrated that the locally accumulated concentrations clearly exceed those required for *in vitro* inhibition of pathogen development (Hahn *et al.*, 1985; Mauch *et al.*, 1988; Snyder and Nicholson, 1990). Likewise, specific inhibitors of PAL have been shown to inhibit phytoalexin or lignin synthesis and to revert incompatible interactions to compatible ones (Waldmüller and Grisebach, 1987; Moerschbacher *et al.*, 1990). The conclusion drawn from such experiments is that in specific plant–pathogen interactions certain defense responses do correlate with the presence of disease resistance genes and thus have the potential to contribute to the resistance phenotype. However, various other studies have failed to establish such a correlation (Schröder *et al.*, 1992; Vogelsang and Barz, 1993; Freytag *et al.*, 1994). Therefore, more conclusive experimental strategies are required to unequivocally resolve this problem.

C. ASSESSMENT OF DEFENSE RESPONSES *IN VIVO*

Undoubtedly, the availability of molecular probes, the development of efficient transformation technologies to introduce foreign genes into plants and pathogens, and the establishment of model plant–pathogen systems that allow facile genetic analyses and manipulations are proving invaluable in identifying important plant determinants for resistance to pathogens.

Direct proof for the role of phytoalexins in disease resistance has been supplied by three types of genetic experiments. First, the degree of virulence of particular pathogen strains was correlated with their ability to degrade plant phytoalexins, such as pisatin in the case of *Nectria haematococca* or rishitin and lubimin in the case of *Gibberella pulicaris* (VanEtten *et al.*, 1989). Secondly, pathogens transformed with genes encoding phytoalexin-detoxifying enzymes became virulent on otherwise non-host plants (see Section IIA). And finally, introduction from grapevine into tobacco of a gene encoding stilbene synthase, the only required enzyme for biosynthesis of the phytoalexin resveratrol, resulted in increased resistance towards infection by the greymold fungus *Botrytis cinerea* (Hain *et al.*, 1993). Whether these plants are also more resistant to other fungal or bacterial pathogens was not reported. Nevertheless, these experiments are the best evidence that phytoalexins can be essential for plant disease resistance.

However, there is also evidence for the contrary. Genetic studies with three *Arabidopsis* phytoalexin-deficient mutants, *pad1*, *pad2*, and *pad3*, argue against a role of phytoalexins in limiting growth of the bacterial pathogen *Pseudomonas syringae* (Glazebrook and Ausubel, 1994). It was demonstrated that the capacity to restrict growth of bacteria carrying defined *avr* genes was not affected in the mutant plants indicating that this form of resistance is not exerted by phytoalexins. All three mutants are unable to synthesize camalexin, the phytoalexin of *Arabidopsis* previously shown to inhibit growth of *Pseudomonas syringae* and the fungus *Cladosporium cucumerinum in vitro* (Tsuji *et al.*, 1992). However, two *pad* mutants did allow enhanced growth of virulent bacteria, suggesting that camalexin may serve to limit growth of these bacteria. Furthermore, the effect of camalexin in restricting fungal pathogen growth was not tested and thus cannot be excluded. Additional results questioning the deterrent functions of phytoalexins were also obtained with *Nectria haematococca*; mutants that have lost the ability to detoxify pisatin nevertheless remained virulent on pea, albeit causing smaller lesions than wild-type fungi (VanEtten *et al.*, 1994).

The importance of products derived from phenylpropanoid pathways in limiting disease development has been suggested from experiments with transgenic tobacco plants having suppressed levels of PAL enzyme activity (Maher *et al.*, 1994). Such plants showed a more rapid and extensive lesion development after infection by the virulent fungal pathogen *Cercospora nicotianae* than wild-type plants.

The role of several PR proteins including PR1, 1,3-β-glucanases, chitinases and osmotin, in limiting pathogen ingress have been experimentally tested both in transgenic plants over-expressing such proteins as well as in plants with antisense-mediated suppression of PR protein synthesis (recently reviewed in Kombrink and Somssich, 1995). The conclusions drawn from such studies strongly suggest that the various types of PR proteins can represent an important part of the general, active defense response that has apparently evolved in all plant species. The results obtained also indicate that the contribution of PR proteins to disease resistance can vary considerably, depending on the plant species and the specific pathogen employed. A proper assessment of PR protein functions in future work will therefore require appropriate bioassays including a broad range of different pathogens.

VI. CONCLUSIONS AND PERSPECTIVES

Enormous progress has been made over the past decade in our understanding of the highly complex molecular events that occur in plant–pathogen interactions. A large body of circumstantial evidence has accumulated implicating diverse biochemical responses in plant defense. Despite species-specific differences, plants do appear to have evolved common basic defense strategies comprising very rapid, localized and systemic defense mechanisms for establishing resistance. Although our knowledge is still rudimentary, a number of common signaling molecules have been identified lending insight to the signal transduction pathways employed for intra- and intercellular communication. Cloning of several pathogen *avr* genes and most recently also of major plant *R* genes should open the way to a better understanding of how cultivar–race specificity is established. Numerous pathogen-inducible genes encoding proteins that may be important for the active plant defense response have also been cloned from various plant species. However, a rigorous demonstration of the role of such specific genes in plant resistance will require a concerted application of biochemical, molecular and genetic approaches. Such studies may be complicated by the fact that the contribution of specific responses to plant defense can vary considerably depending on the plant species or the specific pathogen employed. Additionally, many plant defense-related genes are present in multiple copies within the genome, impeding simple and clean functional correlation, e.g. by "knockout" experiments. Furthermore, the presence of functionally related but serologically distinct pathogen-inducible activities in plants may lead to misinterpretations concerning the relevance of specific enzymes in pathogenesis. This was recently demonstrated for 1,3-β-glucanase-deficient tobacco mutants generated by antisense transformation which apparently compensate for the deficient enzyme activity by producing a functional equivalent (Beffa *et al.*, 1993).

Based on our present knowledge, several strategies have emerged for improving crop resistance to pathogens (outlined in Lamb *et al.*, 1992; Cornelissen and Melchers, 1993; Strittmatter and Wegener, 1993). They include the manipulation of resistance exerted by single genes, multiple gene or regulatory mechanisms. Constitutive expression of single proteins, such as a viral coat protein, chitinase, lysozyme, osmotin, *Bacillus thuringiensis* δ-endotoxin, proteinase inhibitor and α-amylase inhibitor, have already proven successful for enhancing plant resistance in the field. Not surprisingly, the combined expression of different proteins has been found to be synergistic. Manipulation of biosynthetic pathways involving multiple genes, e.g. the biosynthesis of phytoalexins, has not yet been achieved, although it should be feasible to modify rate-limiting steps of pathways or to divert precursors into novel pathways by the introduction of defined genes. Other "two-component-systems", involving cytotoxic compounds such as ribonucleases or HR-inducing *avr* gene products are also being developed (de Wit, 1992; Strittmatter and Wegener, 1993). Ideally, such a strategy must rely on tightly regulated plant promoters which direct expression specifically and exclusively to infection sites. Genetic manipulation of the regulatory mechanisms and signaling processes controlling the coordinate activation of multiple defense responses might be the ultimate approach to modify plant resistance. However, this requires precise knowledge of both the signaling pathways involved and the subsequent metabolic pathways they trigger. Finally, the development of synthetic, non-toxic, chemical inducers of protection that can efficiently induce defense responses offers an attractive alternative to complement the genetic engineering approaches employed to establish plant resistance.

Despite the large advances made in this research field, numerous basic questions related to plant defense remain to be answered. What is the biochemical function of plant *R* genes? What is the nature of the plant receptors capable of perceiving a pathogen and how is the warning signal transduced within the plant cell? What is the molecular basis of the HR and does this process have anything in common with apoptosis, the programmed cell death in animals? How is SAR established? With the molecular and genetic tools at hand it should be possible to address these and other questions related to the physiology and biochemistry of plant defense and to dissect the multicomponent regulatory pathways governing the expression of defense-related genes.

ACKNOWLEDGEMENTS

We thank Drs Klaus Hahlbrock, Wolfgang Knogge and Paul Rushton for critical reading of the manuscript.

REFERENCES

Apostol, I., Heinstein, P. F. and Low, P. S. (1989). Rapid stimulation of an oxidative burst during elicitation of cultured plant cells. *Plant Physiology* **90**, 109–116.

Arlorio, M., Ludwig, A., Boller, T. and Bonfante, P. (1992). Inhibition of fungal growth by plant chitinases and β-1,3-glucanases. A morphological study. *Protoplasma* **171**, 34–43.

Atkinson, M. M. (1993). Molecular mechanisms of pathogen recognition by plants. *Advances in Plant Pathology* **10**, 35–64.

Avdiushko, S. A., Ye, X. S., Hildebrand, D. F. and Kuc, J. (1993). Induction of lipoxygenase activity in immunized cucumber plants. *Physiological and Molecular Plant Pathology* **42**, 83–95.

Baker, C. J., Orlandi, E. W. and Mock, N. M. (1993). Harpin, an elicitor of the hypersensitive response in tobacco caused by *Erwinia amylovora*, elicits active oxygen production in suspension cells. *Plant Physiology* **102**, 1341–1344.

Basse, C. W., Fath, A. and Boller, T. (1993). High affinity binding of a glycopeptide elicitor to tomato cells and microsomal membranes and displacement by specific glycan suppressors. *Journal of Biological Chemistry* **268**, 14724–14731.

Beaulieu, C., Boccara, M. and van Gijsegem, F. (1993). Pathogenic behavior of pectinase-defective *Erwinia chrysanthemi* mutants on different plants. *Molecular Plant–Microbe Interactions* **6**, 197–202.

Beffa, R. S., Neuhaus, J.-M. and Meins, F., Jr (1993). Physiological compensation in antisense transformants: specific induction of an "ersatz" glucan endo-1,3-β-glucosidase in plants infected with necrotizing viruses. *Proceedings of the National Academy of Sciences USA* **90**, 8792–8796.

Bent, A., Kunkel, B., Dahlbeck, D., Brown, K., Schmidt, R., Giraudat, J., Leung, J. and Staskawicz, B. (1994). *RPS2* of *Arabidopsis thaliana*: a leucin-rich repeat class of plant disease resistance gene. *Science* **265**, 1856–1860.

Bergmann, C. W., Ito, Y., Singer, D., Albersheim, P., Darvill, A. G., Benhamou, N., Nuss, L., Salvi, G., Cervone, F. and De Lorenzo, G. (1994). Polygalacturonase-inhibiting protein accumulates in *Phaseolus vulgaris* L. in response to wounding, elicitors and fungal infection. *Plant Journal* **5**, 625–634.

Bohlmann, H. (1994). The role of thionins in plant protection. *Critical Reviews in Plant Sciences* **13**, 1–16.

Boller, T. (1991). Ethylene in pathogenesis and disease resistance. *In* "The Plant Hormone Ethylene" (A. K. Mattoo and J. C. Suttle, eds), pp. 293–314. CRC Press, Boca Raton.

Bowles, D. J. (1990). Defense-related proteins in higher plants. *Annual Review of Biochemistry* **59**, 873–907.

Bradley, D. J., Kjellbom, P. and Lamb, C. J. (1992). Elicitor- and wound-induced oxidative cross-linking of a proline-rich plant cell wall protein: a novel, rapid defense response. *Cell* **70**, 21–30.

Brederode, F. T., Linthorst, H. J. M. and Bol, J. F. (1991). Differential induction of acquired resistance and PR gene expression in tobacco by virus infection, ethephon treatment, UV light and wounding. *Plant Molecular Biology* **17**, 1117–1125.

Cameron, R. K., Dixon, R. A. and Lamb, C. J. (1994). Biologically induced systemic acquired resistance in *Arabidopsis thaliana*. *Plant Journal* **5**, 715–725.

Chen, Z., Ricigliano, J. W. and Klessig, D. F. (1993a). Purification and characterization of a soluble salicylic acid-binding protein from tobacco. *Proceedings of the National Academy of Sciences USA* **90**, 9533–9537.

Chen, Z., Silva, H. and Klessig, D. F. (1993b). Active oxygen species in the induction of plant systemic acquired resistance by salicylic acid. *Science* **262**, 1883–1886.

Collinge, D. B., Gregersen, P. L. and Thordal-Christensen, H. (1994). The induction of gene expression in response to pathogenic microbes. *In* "Mechanisms of Plant Growth and Improved Productivity: Modern Approaches and Perspectives" (A. S. Basra, ed.), pp. 391–433. Marcel Dekker, New York.

Cornelissen, B. J. C. and Melchers, L. S. (1993). Strategies for control of fungal diseases with transgenic plants. *Plant Physiology* **101**, 709–712.

Cutt, J. R. and Klessig, D. F. (1992). Pathogenesis-related proteins. *In* "Genes Involved in Plant Defense" (T. Boller and F. Meins Jr, eds), pp. 209–243. Springer-Verlag, Wien.

da Costa e Silva, O., Klein, L., Schmelzer, E., Trezzini, G. F. and Hahlbrock, K. (1993). BPF-1, a pathogen-induced DNA-binding protein involved in the plant defense response. *Plant Journal* **4**, 125–135.

de Wit, P. J. G. M. (1992). Molecular characterization of gene-for-gene systems in plant–fungus interactions and the application of avirulence genes in control of plant pathogens. *Annual Review of Phytopathology* **30**, 391–418.

Devlin, W. S. and Gustine, D. L. (1992). Involvement of the oxidative burst in phytoalexin accumulation and the hypersensitive reaction. *Plant Physiology* **100**, 1189–1195.

Dickman, M. B., Podila, G. K. and Kolattukudy, P. E. (1989). Insertion of cutinase gene into a wound pathogen enables it to infect intact host. *Nature* **342**, 446–448.

Dietrich, A., Mayer, J. E. and Hahlbrock, K. (1990). Fungal elicitor triggers rapid, transient and specific protein phosphorylation in parsley cell suspension cultures. *Journal of Biological Chemistry* **265**, 6360–6368.

Dietrich, R. A., Delaney, T. P., Uknes, S. J., Ward, E. R., Ryals, J. A. and Dangl, J. L. (1994). Arabidopsis mutants simulating disease resistance response. **Cell** 77, 565–577.

Dixon, R. A. and Harrison, M. J. (1990). Activation, structure and organization of genes involved in microbial defense in plants. *Advances in Genetics* **28**, 165–234.

Dixon, R. A., Harrison, M. J. and Lamb, C. J. (1994). Early events in the activation of plant defense responses. *Annual Review of Phytopathology* **32**, 479–501.

Ebel, J. and Cosio, E. G. (1994). Elicitors of plant defense responses. *International Review of Cytology* **148**, 1–36.

Enyedi, A. J., Yalpani, N., Silverman, P. and Raskin, I. (1992). Signal molecules in systemic plant resistance to pathogens and pests. *Cell* **70**, 879–886.

Farmer, E. E. and Ryan, C. A. (1992). Octadecanoid precursors of jasmonic acid activate the synthesis of wound-inducible proteinase inhibitors. *The Plant Cell* **4**, 129–134.

Farmer, E. E., Johnson, R. R. and Ryan, C. A. (1992). Regulation of expression of proteinase inhibitor genes by methyl jasmonate and jasmonic acid. *Plant Physiology* **98**, 995–1002.

Felix, G., Regenass, M. and Boller, T. (1993). Specific perception of subnanomolar concentrations of chitin fragments by tomato cells: induction of extracellular alkalinization, changes in protein phosphorylation and establishment of a refractory state. *Plant Journal* **4**, 307–316.

Fournier, J., Pouénat, M.-L., Rickauer, M., Rabinovitch-Chable, H., Rigaud, M. and Esquerre-Tugayé, M.-T. (1994). Purification and characterization of elicitor-induced lipoxygenase in tobacco cells. *Plant Journal* **3**, 63–70.

Freytag, S., Arabatzis, N., Hahlbrock, K. and Schmelzer, E. (1994). Reversible cytoplasmic rearrangements precede wall apposition, hypersensitive cell death and defense-related gene activation in potato/*Phytophthora infestans* interactions. *Planta* **194**,123–135.

Gaffney, T., Friedrich, L., Vernooij, B., Negrotto, D., Nye, G., Uknes, S., Ward,

E., Kessmann, H. and Ryals, J. (1993). Requirement of salicylic acid for the induction of systemic acquired resistance. *Science* **261**, 754–756.

Geerts, A., Feltkamp, D. and Rosahl, S. (1994). Expression of lipoxygenase in wounded tubers of *Solanum tuberosum* L. *Plant Physiology* **105**, 269–277.

Geoffroy, P., Legrand, M. and Fritig, B. (1990). Isolation and characterization of a proteinaceous inhibitor of microbial proteinases induced during the hypersensitive reaction of tobacco to tobacco mosaic virus. *Molecular Plant–Microbe Interactions* **3**, 327–333.

Glazebrook, J. and Ausubel, F. M. (1994). Isolation of phytoalexin-deficient mutants of *Arabidopsis thaliana* and characterization of their interactions with bacterial pathogens. *Proceedings of the National Academy of Sciences USA* **91** 8955–8959.

Godiard, L., Grant, M. R., Dietrich, R. A., Kiedrowski, S. and Dangl, J. L. (1994). Perception and response in plant disease resistance. *Current Opinion in Genetics and Development* **4**, 662–671.

Gough, C. L., Genin, S., Zischek, C. and Boucher, C. A. (1992). *hrp* genes of *Pseudomonas solanacearum* are homologous to pathogenicity determinants of animal pathogenic bacteria and are conserved among plant pathogenic bacteria. *Molecular Plant–Microbe Interactions* **5**, 384–389.

Greenberg, J. T., Guo, A., Klessig, D. F. and Ausubel, F. M. (1994). Programmed cell death in plants: a pathogen-triggered response activated coordinately with multiple defense functions. *Cell* **77**, 551–563.

Gross, P., Julius, C., Schmelzer, E. and Hahlbrock, K. (1993). Translocation of cytoplasm and nucleus to fungal penetration sites is associated with depolymerization of microtubles and defence gene activation in infected, cultured parsley cells. *EMBO Journal* **12**, 1735–1744.

Gundlach, H., Müller, M. J., Kutchan, T. M. and Zenk, M. H. (1992). Jasmonic acid is a signal transducer in elicitor-induced plant cell cultures, *Proceedings of the National Academy of Sciences USA* **89**, 2389–2393.

Hahlbrock, K. and Scheel, D. (1989). Physiology and molecular biology of phenylpropanoid metabolism. *Annual Review of Plant Physiology and Plant Molecular Biology* **40** 347–369.

Hahn, M. G., Bonhoff, A. and Grisebach, H. (1985). Quantitative localization of the phytoalexin glyceollin I in relation to fungal hyphae in soybean roots infected with *Phytophthora megasperma* f. sp. *glycinea*. *Plant Physiology* **77**, 591–601.

Hain, R., Reif, H.-J., Krause, E., Langebartels, R., Kindl, H., Vornam, B., Wiese, W., Schmelzer, E., Schreier, P. H., Stöcker, R. H. and Stenzel, K. (1993). Disease resistance results from foreign phytoalexin expression in a novel plant. *Nature* **361**, 153–156.

Hammerschmidt, R. (1993). The nature and generation of systemic signals induced by pathogens, arthropod herbivores and wounds. *Advances in Plant Pathology* **10**, 307–337.

Jahnen, W. and Hahlbrock, K. (1988). Cellular localization of nonhost resistance reactions of parsley (*Petroselinum crispum*) to fungal infection. *Planta* **173**, 197–204.

Jakobek, J. L. and Lindgren, P. B. (1993). Generalized induction of defense responses in bean is not correlated with the induction of the hypersensitive response. *The Plant Cell* **5**, 49–56.

Johal, G. and Briggs, S. (1992). Reductase activity encoded by the *HM1* disease resistance gene in maize. *Science* **258**, 985–987.

Jones, D. A., Thomas, C. M., Hammond-Kosack, K. E., Balint-Kurti, P. J. and Jones, J. D. G. (1994). Isolation of the tomato *Cf-9* gene for resistance to *Cladosporium fulvum* by transposon tagging. *Science* **266**, 789–793.

Kamoun, S., Klucher, K. M., Coffey, M. D. and Tyler, B. M. (1993). A gene encoding a host-specific elicitor protein of *Phytophthora parasitica*. *Molecular Plant–Microbe Interactions* **6**, 573–581.

Kauss, H. (1987). Some aspects of calcium-dependent regulation in plant metabolism. *Annual Review of Plant Physiology* **38**, 47–72.

Kawalleck, P., Plesch, G., Hahlbrock, K. and Somssich, I. E. (1992). Induction by fungal elicitor of *S*-adenosyl-L-methionine synthetase and *S*-adenosyl-L-homocysteine hydrolase mRNAs in cultured cells and leaves of *Petroselinum crispum*. *Proceedings of the National Academy of Sciences USA* **89**, 4713–4717.

Kawalleck, P., Keller, H., Hahlbrock, K., Scheel, D. and Somssich, I. E. (1993). A pathogen-responsive gene of parsley encodes tyrosine decarboxylase. *Journal of Biological Chemistry* **268**, 2189–2194.

Kawalleck, P., Schmelzer, E., Hahlbrock, K. and Somssich, I. E. (1995). Two pathogen-responsive parsley genes encode tyrosine-rich hydroxyproline-rich glycoprotein (HRGP) and an anionic peroxidase. *Molecular and General Genetics* **247**, 444–452.

Keen, N. T. (1990). Gene-for-gene complementarity in plant–pathogen interactions. *Annual Review of Genetics* **24**, 447–463.

Keen, N. T. (1992). The molecular biology of disease resistance. *Plant Molecular Biology* **19**, 109–122.

Kelemu, S. and Collmer, A. (1993). *Erwinia chrysanthemi* EC16 produces a second set of plant-inducible pectate lyase isozymes. *Applied and Environmental Microbiology* **59**, 1756–1761.

Kessmann, H., Staub, T., Hofmann, C., Maetzke, T., Herzog, J., Ward, E., Uknes, S. and Ryals, J. (1994). Induction of systemic acquired resistance in plants by chemicals. *Annual Review of Phytopathology* **32**, 439–459.

Knogge, W. (1991). Plant resistance genes for fungal pathogen — Physiological models and identification in cereal crops. *Zeitschrift für Naturforschung* **46c**, 969–981.

Knogge, W., Gierlich, A., Hermann, H., Wernert, P. and Rohe, M. (1994). Molecular identification and characterization of the *nip1* gene, an avirulence gene from the barley pathogen *Rhynchosporium secalis*. *In* "Advances in Molecular Genetics of Plant–Microbe Interactions, Vol. 3" (M. J. Daniels, J. A. Downie and A. E. Osbourn, eds), pp. 207–214. Kluwer Academic Publishers, Dordrecht.

Kolattukudy, P. E. (1980). Cutin, suberin and waxes. *In* "The Biochemistry of Plants. A Comprehensive Treatise. Vol. 4. Lipids: Structure and Function" (P. K. Stumpf, ed.), pp. 571–645. Academic Press, New York.

Kombrink, E. and Somssich, I. E. (1995). Pathogenesis-related proteins and plant defense. *In* "Plant Relationships" (G. Carroll and P. Tudzynski, eds), The Mycota, **Vol. 6**, Springer-Verlag, Berlin in press.

Kombrink, E., Beerhues, L., Garcia-Garcia, F., Hahlbrock, K., Müller, M., Schröder, M., Witte, B. and Schmelzer, E. (1993). Expression patterns of defense-related genes in infected and uninfected plants. *In* "Mechanisms of Plant Defense Responses" (B. Fritig and M. Legrand, eds), Developments in Plant Pathology, Vol. 2, pp. 236–249. Kluwer Academic Publishers, Dordrecht.

Korfhage, U., Trezzini, G. F., Meier, I., Hahlbrock, K. and Somssich, I. E. (1994). Plant homeodomain protein involved in transcriptional regulation of a pathogen defense-related gene. *The Plant Cell* **6**, 695–708.

Lamb, C. J., Ryals, J. A., Ward, E. R. and Dixon, R. A. (1992). Emerging strategies for enhancing crop resistance to microbial pathogens. *Bio/Technology* **10**, 1436–1445.

Lawrence, G. J., Ellis, J. G. and Finnegan, E. J. (1994). Cloning a rust-resistance gene in flax. *In* "Advances in Molecular Genetics of Plant–Microbe Interactions,

Vol. 3" (M. J. Daniels, J. A. Downie and A. E. Osbourn, eds), pp. 303-306. Kluwer Academic Publishers, Dordrecht.

Lawton, K. A., Beck, J., Potter, S., Ward, E. and Ryals, J. (1994a). Regulation of cucumber class III chitinase gene expression. *Molecular Plant-Microbe Interactions* **7**:48-57.

Lawton, K. A., Potter, S. L., Uknes, S. and Ryals, J. (1994b). Acquired resistance signal transduction in *Arabidopsis* is ethylene independent. *The Plant Cell* **6**, 581-588.

Legendre, L., Rueter, S., Heinstein, P. F. and Low, P. S. (1993). Characterization of the oligogalacturonide-induced oxidative burst in cultured soybean (*Glycine max*) cells. *Plant Physiology* **102**, 233-240.

Levine, A., Tenhaken, R., Dixon, R. and Lamb, C. (1994). H_2O_2 from the oxidative burst orchestrates the plant hypersensitive disease resistance response. *Cell* **79**, 583-593.

Levings, C. S., III and Siedow, J. N. (1992). Molecular basis of disease susceptibility in the Texas cytoplasm of maize. *Plant Molecular Biology* **19**, 135-147.

Linthorst, H. J. M., Brederode, F. T., van der Does, C. and Bol, J. F. (1993). Tobacco proteinase inhibitor I genes are locally, but not systemically induced by stress. *Plant Molecular Biology* **21**, 985-992.

Madamanchi, N. R. and Kuc, J. (1991). Induced systemic resistance in plants. *In* "The Fungal Spore and Disease Initiation in Plants and Animals" (G. T. Cole and H. C. Hoch, eds), pp. 347-362. Plenum, New York.

Maher, E. A., Rate, N. J., Ni, W., Elkind, Y., Dixon, R. A. and Lamb, C. J. (1994). Increased disease susceptibility of transgenic tobacco plants with suppressed levels of preformed phenylpropanoid products. *Proceedings of the National Academy of Sciences USA* **91**, 7802-7806.

Malamy, J. and Klessig, D. F. (1992). Salicylic acid and plant disease resistance. *Plant Journal* **2**, 643-654.

Marmeisse, R., van den Ackerveken, G. F. J. M., Goosen, T. and de Wit, P. J. G. M. (1993). Disruption of the avirulence gene *avr9* in two races of the tomato pathogen *Cladosporium fulvum* causes virulence on tomato genotypes with the complementary resistance gene *Cf9*. *Molecular Plant-Microbe Interactions* **6**, 412-417.

Martin, G. B., Brommenschenkel, S. H., Chunwongse, J., Frary, A., Ganal, M. W., Spivey, R., Wu, T., Earle, E. D. and Tanksley, S. D. (1993). Map-based cloning of a protein kinase gene conferring disease resistance in tomato. *Science* **262**, 1431-1436.

Mauch, F., Mauch-Mani, B. and Boller, T. (1988). Antifungal hydrolases in pea tissue. II. Inhibition of fungal growth by combinations of chitinase and 1,3-β-glucanase. *Plant Physiology* **88**, 936-942.

McGurl, B., Orozco-Cárdenas, M., Pearce, G. and Ryan, C. A. (1994). Overexpression of the prosystemin gene in transgenic tomato plants generates a systemic signal that constitutively induces proteinase inhibitor synthesis. *Proceedings of the National Academy of Sciences USA* **91**, 9799-9802.

Meeley, R. B., Johal, G. S., Briggs, S. P. and Walton, J. D. (1992). A biochemical phenotype for a disease resistance gene of maize. *The Plant Cell* **4**, 71-77.

Mehdy, M. C. (1994). Active oxygen species in plant defense against pathogens. *Plant Physiology* **105**, 467-472.

Meier, B. M., Shaw, N. and Slusarenko, A. J. (1993). Spatial and temporal accumulation of defense gene transcripts in bean (*Phaseolus vulgaris*) leaves in relation to bacteria-induced hypersensitive cell death. *Molecular Plant Microbe Interactions* **6**, 453-466.

Melan, M. A., Dong, X., Endara, M. E., Davis, K. R., Ausubel, F. M. and Peterman,

T. K. (1993). An *Arabidopsis thaliana* lipoxygenase gene can be induced by pathogens, abscisic acid and methyl jasmonate. *Plant Physiology* **101**, 441–450.

Mindrinos, M., Katagiri, F., Yu, G.-L. and Ausubel, F. M. (1994). The *A. thaliana* disease resistance gene *RPS2* encodes a protein containing a nucleotide-binding site and leucine-rich repeats. *Cell* **78**, 1089–1099.

Moerschbacher, B. M., Noll, U., Gorrichon, L. and Reisener, H.-J. (1990). Specific inhibition of lignification breaks hypersensitive resistance of wheat to stem rust. *Plant Physiology* **93**, 465–470.

Mueller, M. J., Brodschelm, W., Spannagl, E. and Zenk, M. H. (1993). Signaling in the elicitation process is mediated through the octadecanoid pathway leading to jasmonic acid. *Proceedings of the National Academy of Sciences USA* **90**, 7490–7494.

Narváez-Vásquez, J., Pearce, G., Orozco-Cárdenas, M. L., Franceschi, V. R. and Ryan, C. A. (1995). Autoradiographic and biochemical evidence for the systemic translocation of systemin in tomato plants. *Planta* **195**, 593–600.

Nicholson, R. L. and Hammerschmidt, R. (1992). Phenolic compounds and their role in disease resistance. *Annual Review of Phytopathology* **30**, 369–389.

Nürnberger, T., Nennstiel, D., Jabs, T., Sacks, W. R., Hahlbrock, K. and Scheel, D. (1994). High affinity binding of a fungal oligopeptide elicitor to parsley plasma membranes triggers multiple defenses responses. *Cell* **78**, 449–460.

Oeser, B. and Yoder, O. C. (1994). Pathogenesis by *Cochliobolus heterostrophus* transformants expressing a cutinase-encoding gene from *Nectria haematococca*. *Molecular Plant–Microbe Interactions* **7**, 282–288.

Orozco-Cardenas, M., McGurl, B. and Ryan, C. A. (1993). Expression of an antisense prosystemin gene in tomato plants reduces resistance toward *Manduca sexta* larvae. *Proceedings of the National Academy of Sciences USA* **90**, 8273–8276.

Pautot, V., Holzer, F. M. and Walling, L. L. (1991). Differential expression of tomato proteinase inhibitor I and II genes during bacterial pathogen invasion and wounding. *Molecular Plant–Microbe Interactions* **4**, 284–292.

Pontier, D., Godiard, L., Marco, I. and Roby, D. (1994). *hsr203J*, a tobacco gene whose activation is rapid, highly localized and specific for incompatible plant/pathogen interactions. *Plant Journal* **5**, 507–521.

Pryor, T. and Ellis, J. (1993). The genetic complexity of fungal resistance genes in plants. *Advances in Plant Pathology* **10**, 281–305.

Roby, D., Toppan, A. and Esquerré-Tugayé, M.-T. (1987). Cell surfaces in plant micro-organism interactions. VIII. Increased proteinase inhibitor activity in melon plants in response to infection by *Colletotrichum lagenarium* or to treatment with an elicitor fraction from this fungus. *Physiological and Molecular Plant Pathology* **30**, 453–460.

Rogers, L. M., Flaishman, M. A. and Kolattukudy, P. E. (1994). Cutinase gene disruption in *Fusarium solani* f sp *pisi* decreases its virulence on pea. *The Plant Cell* **6**, 935–945.

Ryan, C. A. (1990). Protease inhibitors in plants: genes for improving defense against insects and pathogens. *Annual Review of Phytopathology* **28**, 425–449.

Ryan, C. A. (1992). The search for the proteinase inhibitor-inducing factor, PIIF. *Plant Molecular Biology* **19**, 123–133.

Ryan, C. A. and Farmer, E. E. (1991). Oligosaccharide signals in plants: a current assessment. *Annual Review of Plant Physiology and Plant Molecular Biology* **42**, 651–674.

Salmond, G. P. C. (1994). Secretion of extracellular virulence factors by plant pathogenic bacteria. *Annual Review of Phytopathology* **32**, 181–200.

Schäfer, W. (1994). Molecular mechanisms of fungal pathogenicity to plants. *Annual Review of Phytopathology* **32**, 461–477.

Schäfer, W., Straney, D., Ciuffetti, L., Van Etten, H. D. and Yoder, O. C. (1989). One enzyme makes a fungal pathogen, but not a saprophyte, virulent on a new host plant. *Science* **246**, 247–249.

Schmelzer, E., Krüger-Lebus, S. and Hahlbrock, K. (1989). Temporal and spatial patterns of gene expression around sites of attempted fungal infection in parsley leaves. *The Plant Cell* **1**, 993–1001.

Schröder, M., Hahlbrock, K. and Kombrink, E. (1992). Temporal and spatial patterns of 1,3-β-glucanase and chitinase induction in potato leaves infected by *Phytophthora infestans*. *Plant Journal* **2**, 161–172.

Schumann, G. L. (1991). "Plant Diseases: Their Biology and Social Impact". American Phytopathological Society, St Paul, Minnesota.

Scott-Craig, J. S., Panaccione, D. G., Cervone, F. and Walton, J. D. (1990). Endopolygalacturonase is not required for pathogenicity of *Cochliobolus carbonum* on maize. *The Plant Cell* **2**, 1191–1200.

Sela-Buurlage, M. B., Ponstein, A. S., Bres-Vloemans, S. A., Melchers, L. S., van den Elzen, P. J. M. and Comelissen, B. J. C. (1993). Only specific tobacco (*Nicotiana tabacum*) chitinases and β-1,3-glucanases exhibit antifungal activity. *Plant Physiology* **101**, 857–863.

Sharon, A., Fuchs, Y. and Anderson, J. D. (1993). The elicitation of ethylene biosynthesis by a *Trichoderma* xylanase is not related to the cell wall degradation activity of the enzyme. *Plant Physiology* **102**, 1325–1329.

Shibuya, N., Kaku, H., Kuchitsu, K. and Maliarik, M. J. (1993). Identification of a novel high-affinity binding site for *N*-acetylchitooligosaccharide elicitor in the membrane fraction from suspension cultured rice cells. *FEBS Letters* **329**, 75–78.

Silverman, P., Nuckles, E., Ye, Y. S., Kuc, J. and Raskin, I. (1993). Salicylic acid, ethylene and pathogen resistance in tobacco. *Molecular Plant–Microbe Interactions* **6**, 775–781.

Snyder, B. A. and Nicholson, R. L. (1990). Synthesis of phytoalexins in sorgum as a site-specific response to fungal ingress. *Science* **248**, 1637–1639.

Somssich, I. E. (1994). Regulatory elements governing pathogenesis-related (PR) gene expression. *In* "Results and Problems in Cell Differentiation" (L. Nover, ed.), **Vol. 20**, pp. 163–179. Springer-Verlag, Berlin, Heidelberg.

Somssich, I. E., Bollmann, J., Hahlbrock, K., Kombrink, E. and Schulz, W. (1989). Differential early activation of defense-related genes in elicitor-treated parsley cells. *Plant Molecular Biology* **12**, 227–234.

Stahl, D. J. and Schäfer, W. (1992). Cutinase is not required for fungal pathogenicity on pea. *The Plant Cell* **4**, 621–629.

Stirpe, F., Barbieri, L., Battelli, M. G., Soria, M. and Lappi, D. A. (1992). Ribosome-inactivating proteins from plants: present status and future prospects. *Bio/ Technology* **10**, 405–412.

Strittmatter, G. and Wegener, D. (1993). Genetic engineering of disease and pest resistance in plants: present state of the art. *Zeitschrift für Naturforschung* **48c**, 673–688.

Taylor, J. L., Fritzemeier, K.-H., Häuser, I., Kombrink, E., Rohwer, F., Schröder, M., Strittmatter, G. and Hahlbrock, K. (1990). Structural analysis and activation by fungal infection of a gene encoding a pathogenesis-related protein in potato. *Molecular Plant–Microbe Interactions* **3**, 72–77.

Tsuji, J., Jackson, E. P., Gage, D. A., Hammerschmidt, R. and Somerville, S. C. (1992). Phytoalexin accumulation in *Arabidopsis thaliana* during the hypersensitive reaction to *Pseudomonas syringae* pv *syringae*. *Plant Physiology* **98**, 1304–1309.

Tzeng, D. D. and DeVay, J. E. (1993). Role of oxygen radicals in plant disease development. *Advances in Plant Pathology* **10**, 1–34.

Uknes, S., Mauch-Mani, B., Moyer, M., Potter, S., Williams, S., Dincher, S., Chandler, D., Slusarenko, A., Ward, E. and Ryals, J. (1992). Acquired resistance in *Arabidopsis*. *The Plant Cell* **4**, 645-656.

Uknes, S., Winter, A. M., Delaney, T., Vernooij, B., Morse, A., Friedrich, L., Nye, G., Potter, S., Ward, E. and Ryals, J. (1993). Biological induction of systemic acquired resistance in *Arabidopsis*. *Molecular Plant-Microbe Interactions* **6**, 692-698.

VanEtten, H. D., Matthews, D. E. and Matthews, P. S. (1989). Phytoalexin detoxification: importance for pathogenicity and practical implications. *Annual Review of Phytopathology* **27**, 143-164.

VanEtten, H., Soby, S., Wasmann, C. and McCluskey, K. (1994). Pathogenicity genes in fungi. *In* "Advances in Molecular Genetics of Plant-Microbe Interactions, Vol. 3" (M. J. Daniels, J. A. Downie and A. E. Osbourn, eds), pp. 161-170. Kluwer Academic Publishers, Dordrecht.

van Gijsegem, F., Somssich, I. E. and Scheel, D. (1995). Activation of defense-related genes in parsley leaves by infection with *Erwinia chrysanthemi*. *European Journal of Plant Pathology* in press.

Vernooij, B., Friedrich, L., Morse, A., Reist, R., Kolditz-Jawhar, R., Ward, E., Uknes, S., Kessmann, H. and Ryals, J. (1994). Salicylic acid is not the translocated signal responsible for inducing systemic acquired resistance but is required in signal transduction. *The Plant Cell* **6**, 959-965.

Viard, M.-P., Martin, F., Pugin, M., Ricci, P. and Blein, J.-P. (1994). Protein phosphorylation is induced in tobacco cells by the elicitor cryptogein. *Plant Physiology* **104**, 1245-1249.

Vigers, A. J., Wiedemann, S., Roberts, W. K., Legrand, M., Selitrennikoff, C. P. and Fritig, B. (1992). Thaumatin-like pathogenesis-related proteins are antifungal. *Plant Science* **83**, 155-161.

Vogelsang, R. and Barz, W. (1993). Purification, characterization and differential hormonal regulation of a β-1,3-glucanase and two chitinases from chickpea (*Cicer arietinum* L.). *Planta* **189**, 60-69.

Waldmüller, T. and Grisebach, H. (1987). Effects of R-(1-amino-2-phenylethyl) phosphonic acid on glyceollin accumulation and expression of resistance to *Phytopthora megasperma* f.sp. *glycinea* in soybean. *Planta* **172**, 424-430.

Walton, J. D. (1994). Deconstructing the cell wall. *Plant Physiology* **104**, 1113-1118.

Ward, E. R., Uknes, S. J., Williams, S. C., Dincher, S. S., Wiederhold, D. L., Alexander, D. C., Ahl-Goy, P., Métraux, J.-P. and Ryals, J. A. (1991). Coordinate gene activity in response to agents that induce systemic acquired resistance. *The Plant Cell* **3**, 1085-1094.

Wei, Z.-M., Laby, R. J., Zumoff, C. H., Bauer, D. W., He, S. Y., Collmer, A. and Beer, S. (1992). Harpin, elicitor of the hypersensitive response produced by the plant pathogen *Erwinia amylovora*. *Science* **257**, 85-88.

Whiteham, S., Dinesh-Kumar, S. P., Choi, D., Hehl, R., Corr, C. and Baker, B. (1994). The product of the tobacco mosaic virus resistance gene *N*: similarity to toll and the interleukin-1 receptor. *Cell* **78**, 1101-1115.

Wildon, D. C., Thain, J. F., Minchin, P. E. H., Gubb, I. R., Reilly, A. J., Skipper, Y. D., Doherty, H. M., O'Donnell, P. J. and Bowles, D. J. (1992). Electric signalling and systemic proteinase inhibitor induction in the wounded plant. *Nature* **360**, 62-65.

Willis, D. K., Rich, J. J. and Hrabak, E. M. (1991). *hrp* genes of phytopathogenic bacteria. *Molecular Plant-Microbe Interactions* **4**, 132-138.

Wolter, M., Hollricher, K., Salamini, F. and Schulze-Lefert, P. (1993). The *mlo* resistance alleles to powdery mildew infection in barley trigger a developmentally controlled defence mimic phenotype. *Molecular and General Genetics* **239**, 122-128.

Xu, Y., Chang, P.-F. L., Liu, D., Narasimhan, M. L., Raghothama, K. G., Hasegawa, P. M. and Bressan, R. A. (1994). Plant defense genes are synergistically induced by ethylene and methyl jasmonate. *The Plant Cell* **6**, 1077–1085.

Yalpani, N., Shulaev, V. and Raskin, I. (1993). Endogenous salicylic acid levels correlate with accumulation of pathogenesis-related proteins and virus resistance in tobacco. *Phytopathology* **83**, 702–708.

Yalpani, N., Enyedi, A. J., León, J. and Raskin, I. (1994). Ultraviolet light and ozone stimulate accumulation of salicylic acid, pathogenesis-related proteins and virus resistance in tobacco. *Planta* **193**, 372–376.

On the Nature and Genetic Basis for Resistance and Tolerance to Fungal Wilt Diseases of Plants

C. H. BECKMAN and E. M. ROBERTS

Department of Plant Sciences, University of Rhode Island, Kingston, Rhode Island 02881 USA

Advances in Botanical Research Vol. 21
incorporating Advances in Plant Pathology
ISBN 0–12–005921–5

I. INTRODUCTION

Wilt diseases, beginning with Fusarium wilt of tomato, were first reported 100 years ago and, because of their worldwide significance on many crops, have been studied intensively ever since. A monograph by Walker (1971) covered much of the literature on Fusarium wilt of tomato through 1970. A subsequent monograph by Beckman (1987) covered the *Fusarium*/tomato literature through the next 15 years and broadened the scope to include Fusarium and Verticillium wilts of other crop plants. The ultimate purpose of the latter publication was to begin the building of models. The first model was to portray our understanding of the life cycle of the pathogen in relation to its survival and the presentation of inoculum to infection model of the host. The second was to portray our understanding of the host–parasite interactions, and connect all the relevant findings with respect to disease resistance or susceptibility and the causation of disease within a framework of time and space.

Seven years have passed since the latter monograph was published. The models, though essentially sound, are in need of updating. Furthermore, the amount of literature dealing with molecular genetics was scant at the time and received only passing mention. It is the purpose of this chapter to amend the models in the light of recent findings and to deal specifically with molecular studies. Because of space constraints, much of the background information and citations will be omitted and the reader will often be referred to earlier monographs which will, in turn, cite original papers.

II. UPDATING THE MODELS

A. AMENDING THE PATHOGEN/DISEASE CYCLE MODEL

A model of the host–parasite interactions in wilt disease was originally proposed by Talboys to depict two functional stages of disease development. These were the determinative phase, in which the extent of colonization of the vascular system of the host by the pathogen was determined, and the expressive phase, in which symptoms developed. This model was expanded by Beckman (1987) to include the saprophytic phase of the life cycle of a pathogen, as represented by *Fusarium oxysporum*. That model has been modified in several ways (Fig. 1).

1. Survival of the pathogen

First, all soil-borne, fungal, vascular parasites must, during the saprophytic phase of their life cycles, survive for months, and sometimes years, without the presence of their preferred hosts. They produce resting structures of some sort (chlamydospores, microsclerotia, or altered mycelium) on roots or within

PATHOGEN LIFE CYCLE / DISEASE CYCLE

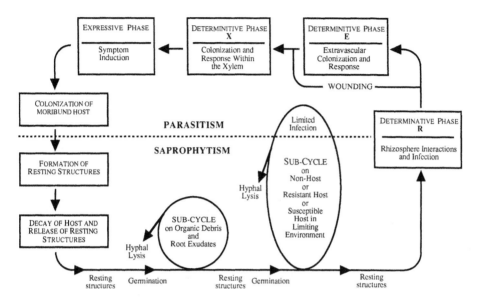

Fig. 1.   The life cycle of soil-borne, wilt-causing fungi, including their saprophytic and parasitic growth and successive phases of colonization and pathogenesis.

various moribund hosts that enable them to survive for extended periods of time. Thus the overall model was altered editorially to broaden the concept to various pathogens; in this context, the term resting structures, rather than "chlamydospores" is appropriate.

These resting structures are readily stimulated into rapid germination and growth by nutrients that are released from the extending roots of a variety of plants (preferred hosts or not), especially at the elongating region of root tips (Huisman and Gerik, 1989), or are stimulated by nutrients that diffuse from organic debris. The rate of root growth through the soil and the limited zone of effective exudates near the root tip, however, dictate that the processes of germination of fungal resting structures, growth of germ tubes and achieving contact with host roots must be rapid. As the root at the point of colonization matures and the level of exudates declines, most of these organisms again revert to resting structures.

Fusarium and *Verticillium*, however, can also make incursions into the epidermal and cortical cells of host and non-host plants before cycling back into resting structures (Fig. 1). In this manner pathogenic and non-pathogenic forms can maintain themselves indefinitely in the presence or absence of the preferred host (Gordon *et al.*, 1989). Whether a plant is susceptible or immune to systemic vascular invasion appears to be unrelated to the ability of the fungus to colonize the root cortex (Huisman and Gerik, 1989). Apparently there is a complex of root flora, in immediate contact with all root surfaces or within root tissues, that is readily available should an opportunity to invade the vascular tissues, by whatever means, present itself.

2. Determinative phases of the host–parasite interaction

This complex of root flora is the basis for a second modification to the overall model of the pathogen/disease cycle (Fig. 1) at the transition from the sapro-phytic to the parasitic phase of the pathogen. Even when the pathogen is pre-sent in the soil, it is now apparent that other microflora among the soil biota can determine whether disease develops or not in otherwise susceptible plants (Alabouvette, 1989; Louvet, 1989). This determining effect is demonstrated in certain regions having soils designated as suppressive, as opposed to others in which the soils are designated as conducive to disease. The suppressive character is clearly biological in nature and dependent upon soil characteristics that allow certain organisms to compete effectively in a mixed biota. Thus, in the modified model, the determinative effect of soil type on the outcome of infection is included in an expanded determinative phase (DP) and noted as *DP-R* to represent the influence of the rhizosphere environment. We will return later to the question of how conducive and suppressive soils may deter-mine the disease outcome.

A third modification of the model concerns the mode of entry of the patho-gen into the xylem elements. There are numerous reports of direct penetration of the pathogen through root cap and cortical tissues into newly formed xylem as indicated in the model (Fig. 1). In most cases, *Verticillium* apparently can enter directly. With *Fusarium*, however, entry appears often to be associated with wounding, and sometimes occurs through senescing tissues (Beckman, 1987). We have therefore added wounding to the model (Fig. 1) as a by-pass mechanism of entry into the xylem elements. This method of entry is likely to occur because of the high populations of selected flora, including patho-gens, that normally inhabit epidermal and cortical tissues of plants and the large number of root feeding fauna in the soil.

We have further modified the model to deal with the processes by which plants defend themselves and those by which pathogenic organisms overcome such defenses in *DP-E*, the determinative phase of disease within extravascular tissues, and in *DP-X*, the determinative phase within xylem tissues. The pro-cesses within *DP-X* will form a core of our understanding of wilt diseases caused by fungi. Ultimately, we will then build upon that fundamental under-standing to explore the genetic basis for resistance and susceptibility and for the relative suppressiveness and conduciveness of soils.

B. UPDATING THE MODEL OF DETERMINATIVE PHASE-E

Talboys (see Beckman, 1987) showed a cellular response in root epidermal and cortical cells of hops following infection with *Verticillium albo-atrum* that resulted in the formation of lignitubers. When successfully developed, these lignitubers apparently prevented further penetration by the parasite into host tissues. The formation of papillae, including the deposition of callose-containing apposition layers and subsequent lignification, is a ubiquitous response of plant cells to attempted penetration by parasites (see Beckman, 1987). The rates of deposition and subsequent lignification determine the degree to which cells are successful in preventing invasion. Jordan *et al.* (1988) have shown that a highly significant reduction in the rate of invasion of epidermal, cortical, and xylem parenchyma cells by *F.o.* f. sp. *apii* in a moderately resistant cultivar of celeriac was associated with a stronger callose response than in highly suscepti- ble cultivars of celery and celeriac. No lignification response was detected, however, within 24 h after inoculation.

If successful in ramifying through cortical tissues of hops, advancing *Verticillium* hyphae eventually encounter the endodermis, where the infection is generally halted (see Beckman, 1987). It is interesting to speculate why this is so. In cotton, and we suspect in plants generally, the endodermal cells uniformly contain phenolics that are synthesized during differentiation and stored in a reduced state. These phenolics are readily released by any stimulus that causes decompartmentation. They then encounter polyphenoloxidase and peroxidase which are localized in plastids and cell walls, respectively. The oxidized phenolics are highly reactive and toxic, polymerizing with many host cell and parasite structures and their enzyme systems, thus disrupting their function. In effect, they "lignify" any structures they meet. *Verticillium* can occasionally penetrate the endodermis. It does so by forming massive ropey structures (composed of numerous, fused hyphal strands) that serve to penetrate through the endodermis (see Beckman, 1987) after which entry into vascular elements readily occurs.

C. UPDATING THE MODEL OF DETERMINATIVE PHASE-X

Once having entered the xylem elements, whether by direct penetration, by means of senescent tissues, or by means of wounds, a parasite encounters an extensive avenue of distribution. The evidence indicates that hyphal growth upward through the vascular system is many times too slow to account for the demonstrated rate of distribution of vascular pathogens in the xylem of any susceptible host, large or small. In a large oak tree the rate of hyphal growth would have to be about 30 cm per day, in a tomato plant about 2 cm per day, far greater than the actual rate of growth in vessels. To take advant- age of this vascular avenue, the pathogen must shift its growth habit from a tissue-ramifying, hyphal mode to one of rapid germination and production of successive generations of spores that can leap-frog upward from one vessel ending to the next in the transpiration stream. In this manner the plant could be colonized systemically, unless the plant can by some means contain the invasion within the initially penetrated vessels, or at least retard its progress. It is this complex process of defense that we shall now explore and seek to model.

1. Defining the time–space boundaries of the model

The xylem elements of plants, even in secondary xylem tissues where vessels are large, have a finite length that varies from a few centimeters to 30 cm in banana, or even a meter in deciduous trees. The vessel length, and thus the free passage distance of fungal spores in the transpiration stream, is determined by the frequency with which gridded or latticed end-walls are formed during normal differentiation. These end walls allow relatively free passage of the sap stream, but serve to screen out fungal spores (Fig. 2). Thus the average free passage distance of fungal spores within a given vascular element ranges from a few centimeters to a meter, depending upon the host plant species. To regain distribution potential in the next vessel element above, the fungus must germinate, penetrate the end-wall and produce a mature conidium that can then be carried in the transpiration stream to the next vessel ending above. The rate of vascular colonization indicates that successive spore passages must occur every 2–3 days. That is also the observed rate of conidium formation within the vascular elements of plants. Thus, free passage of successive generations of conidia can occur every 2–3 days in a highly susceptible plant. If the plant is to contain the infection, however, it must do so within that time frame.

Our model of *DP-X* (Fig. 2) has therefore been given time boundaries of 0–7 days. This time frame extends from time-0 (*t-0*), at which time spores are experimentally introduced into severed vessels of roots or hypocotyls, to 7 days (*t-7*), to include all known interactions. Host factors in this interplay, and the resulting localization when host factors are predominant, are shown on the left; those of the pathogen, and failure in the localization process when pathogen factors predominate, are shown on the right. The model also includes the estimated times within which these responses occur. The time frame *t-0* to *t-7* thus includes the initial infection period and extends through the first secondary spore passage, if any, and the ensuing interaction. This time frame also permits us to distinguish relevant events that occur in resistant-type interactions as opposed to those in susceptible-type interactions.

Our model for *DP-X* has also been given relevant space boundaries, again determined by end-walls. When spore suspensions (plus tracer particles) are introduced experimentally into severed vessels, uptake of spores is rapid and within a few minutes the inoculum is continuous up to the first end-wall trapping site. This vascular space, in which the inoculum is initially present, as

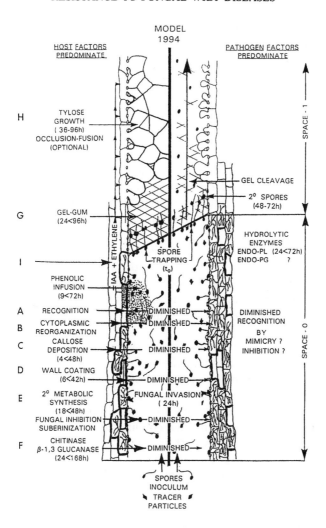

Fig. 2. A time–space model of determinative phase-X in which are entered host–parasite interactions that occur within space-0 (the initially inoculated vessel element, below the vessel ending, and the surrounding vascular parenchyma tissue) and in space-1 (the next vessel, above the inoculated vessel, and its surrounding parenchyma tissue). The left side of the model shows the various defense processes of the host and the times of their occurrence in space-0 and space-1 that, when predominant, serve to inhibit the fungus and to localize the infection. The right side of the model shows the processes of the pathogen and the times of their occurrence in space-0 and space-1 that, when predominant, enable the parasite to escape space-0 and traverse space-1. Inoculum is introduced through severed vessels (bottom of model) and drawn upwards through space-0 by transpirational pull within 10 min (time-0).

well as the xylem parenchyma tissues surrounding it, have been designated space-0, that is, the space that was infested at time 0. The processes of defense that occur within space-0 (i.e., below the initial trapping sites) are, as we shall see, designed to prevent the infection from spreading laterally into the vascular parenchyma cells and adjacent vessels.

The space between the initial trapping site and the second trapping site has been designated space-1. The processes of defense here are much different from those in space-0. They are designed to prevent the secondary fungal spores from advancing through space-1 by a longitudinal leap upward in the transpiration stream. Within space-0 and space-1, different strategies are employed. The refined model presented herein (Fig. 2) has been updated to reflect more clearly this dichotomy of function between space-0 and space-1 and more accurately define the timing of these events.

2. Defining defense-responses in space-0

In all the studies reported in this section, the inoculum was introduced as a spore suspension directly into the xylem elements through severed vessels or needle injections and, in most instances, with an admixture of colored tracer particles to define the extent of initial uptake. Thus, time-0 and space-0 were clearly defined.

It is apparent that as the fungal spores germinate within the vascular elements (some 6 h after inoculation) the germ tubes grow directly toward pits, presumably by a chemotropic response (Beckman, 1987). By 24 h the adjacent contact cells (parenchyma cells in immediate contact with vessels) can be thoroughly colonized. Whether they are or not depends on the rate at which the host cells respond to prevent such colonization. Contact cells, in general, accomplish this task effectively with responses that occur within a period of a few hours to 3 days after inoculation.

Beckman et al. (1991) found a considerable alteration in the cytoplasm of contact cells within 1 h after inoculation of I-gene resistant tomato plants (Fig. 3) in which the cytoplasm on the side of cells immediately adjacent to the infected vessel became vacuolated and distended several times its normal dimension ($P = 0.001$). Such a shift in cytoplasmic structure indicates that cells are becoming more active metabolically (Mauseth, 1988, p. 31). Within 24 h the cytoplasm on the side of the cell distal to the infection also became distended, but to a much lesser degree. The cytoplasm of contact cells in a susceptible near-isoline of tomato also became distended, but to a much lesser degee than in the resistant host ($P = 0.002$) and only after many hours of exposure to the inoculum. These results indicate that a powerful recognition event and response occurs in the resistant interaction where contact between host and pathogen cells has occurred. This recognition has been noted at A in Fig. 2. Recognition in the susceptible host appears to be weaker or delayed, or the response is inhibited. The alteration in the cytoplasm is noted at B in Fig. 2.

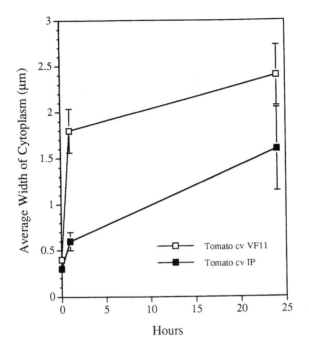

Fig. 3. The average cytoplasmic width (μm) on the proximal side of cells from an inoculated susceptible tomato (near isoline cv IP, closed squares) and a resistant tomato (cv VF, open squares) when measured at 0 h (5 min), 1 h and 24 h after inoculation with *Fusarium oxysporum* f. sp. *lycopersici*. Differences were significant with respect to time ($P = 0.002$) and highly significant with respect to cultivar ($P = 0.001$). Total cell widths ranged from 12 to 22 μm. Reprinted from Beckman *et al.* (1991) with permission.

In other experiments (Mueller *et al.*, 1994) it was shown that within 4–6 h after inoculating a resistant tomato cultivar with *F.o.l.* race 1, a callose-containing apposition layer was synthesized and deposited in contact cells (noted at C in Fig. 2). These deposits occurred initially at the pit walls, but later extended along the entire wall proximal to the infected vessel. They finally took on a marbled appearance with various electron translucent areas and osmiophilic, electron opaque areas. This alteration, it was concluded, represented a lignification or suberization of the apposition layers as was initially indicated by Talboys (see Beckman, 1987). This process is considered later.

In another study (Beckman *et al.*, 1989), the effectiveness of this system of lateral defense was measured in inoculated I-gene-resistant tomato plants and compared with that in inoculated near-isogenic susceptible plants. In both cultivars, 2–15% of the contact cells showed what was called a hypersensitive reaction in which the cytoplasm of contact cells became disorganized and both the cytoplasm and cell walls became strongly osmiophilic, but no cell penetration by the pathogen was observed (Fig. 4). In addition, however, some 40% of the contact cells in the resistant cultivar showed an intense callose deposit whereas less than 8% of the contact cells in the susceptible host showed callose deposits (Fig. 4). None of the cells with callose apposition layers (presumably suberized because of their osmiophilic nature) became infected. The remaining contact cells, 50% in the resistant and 90% in the susceptible plants, were infected and thoroughly colonized.

When the first and second parenchyma cells adjacent to the contact cells were examined, progressively fewer cells were infected (Fig. 4). The difference between the resistant and susceptible hosts, however, was dramatic. Less than 3% of the third layer of cells were infected in the resistant cultivar whereas 70% of these cells in the susceptible host were infected. Thus a system of "defense in depth" was demonstrated which was far more effective in the resistant than in the susceptible host. This defense was largely dependent on the capacity and the rate to which xylem parenchyma cells could deposit callose-containing apposition layers and lignify or suberize them. This difference is noted proportionally by the number of hyphal-invaded cells on the left and right. Fig. 2.

Shi *et al.* (1991b) studied the responses of contact cells in cotton adjacent to *F.o. vasinfectum* infections in space-0 over a period of 0–72 h. Some cells were invaded and their protoplasts rapidly disintegrated within 24 h. The protoplasts in other cells degenerated and became highly osmiophilic, but were not invaded. In a third type of reaction the cells responded strongly with the deposition of an apposition layer that also became strongly osmiophilic. These cells apparently remained healthy and invasion by the pathogen was never observed. Similar responses occurred in both resistant and susceptible cultivars but the resistant reactions (types 2 and 3) were faster and more pronounced in the resistant cultivar.

Shi *et al.* (1991a) also found osmiophilic droplets to be formed initially in cisternae of the endoplasmic reticulum (ER) and membranous envelopes of mitochondria, and occasionally in membrane envelopes of plastids and nuclei. Developmental evidence suggested that this material was excreted into and combined with the apposition layers and also excreted into the vessel lumina of space-0.

Fig. 4.   The percentage colonization of contact parenchyma cells and of first (1°) adjacent and second (2°) adjacent cells of the vascular parenchyma tissues of (A) a susceptible cultivar or (B) a resistant cultivar of tomato following infection with *Fusarium oxysporum* f. sp. *lycopersici* race 1. Note that none of the cells showing callose deposition or a hypersensitive reaction (HR) were infected, and that such cells protected the immediately adjacent cells from challenge by the fungus. The fractions represent the number of cells that showed a particular response or condition divided by the total number of cells counted. Reprinted from Beckman *et al.* (1989), with permission.

SUSCEPTIBLE CULTIVAR

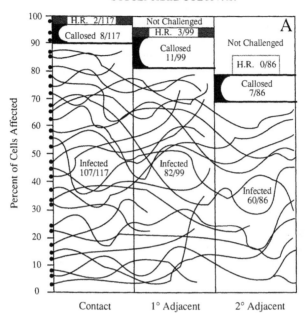

RESISTANT CULTIVAR

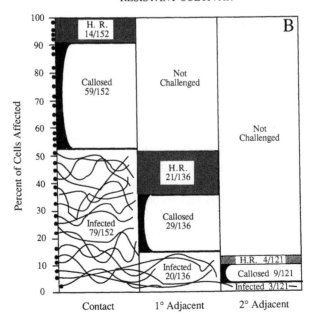

A related facet of this lateral defense had been documented in a series of papers by Robb and various colleagues (see Robb *et al.*, 1991) in which they reported wall coatings in vessels and surrounding parenchyma cells of tomato plants inoculated with *Verticillium albo-atrum*. By using alternate thick and thin sections of plastic-embedded tissue, they were able to integrate information from histochemical and structural studies at both the light and electron microscope levels of magnification. They discovered two types of wall coatings. The first (type A) was lignified and associated with cells without secondary wall deposits, whereas the second (type B) was suberized and associated with walls having secondary thickenings. These studies were extended to highly resistant, moderately resistant and susceptible clones of alfalfa inoculated with *Verticillium albo-atrum*. A significant inverse relationship was found between the levels of coatings and the frequency of pathogen penetration at pit membranes 24 h after inoculation. It was concluded that these wall coatings provided an effective barrier to fungal penetration and, thus, contributed to the overall resistance process. Newcombe and Robb (1989) expanded on the histochemical studies to show the accumulation and redistribution of lipid bodies in contact cells adjacent to conidial trapping sites in alfalfa plants inoculated with *Verticillium*. This lipid coating then spread to the vessel walls. The coating, which showed a suberin reaction, was visible by 9 h after inoculation, and increased to a maximum at 24 h in a resistant cultivar. The response was somewhat slower in a susceptible plant clone. Host wall coating is noted at D in Fig. 2.

Robb *et al.* (1991) then used chemical depolymerization and combined gas-liquid chromatogaphy–mass spectrometry to analyze the contents of the wall coating. The presence of alkane -α,-γ-diols confirmed the suberin nature of the wall coatings and showed quantitative differences that correlated with histological differences, e.g., higher levels in resistant, lower in susceptible plants. Using Northern blot analysis, they also showed corresponding differences in the levels of anionic peroxidase mRNA, localized in responding vascular bundles, that would be essential for polymerizing the aromatic constituents of suberin. This secondary metabolism is noted at E in Fig. 2.

Newcombe *et al.* (1989) inoculated numerous plants from four alfalfa cultivars with *Va-a*, selected nine of these showing a range of symptom expression, and propagated these plants vegetatively from apparently healthy shoots. They reinoculated the plants and determined the number of initially infected vessels (those with tracer particles) and secondarily infected vessels (those with fungus, but no particles). They then calculated a colonization ratio (CR) by dividing the latter by the former. This ratio, then, provided a measure of the rate of lateral spread of the organism, from vessel to abutting vessel, within space-0 of the host. They found a range of CR values (from 0.9 to 12.6) and proposed that the CR value would be a useful tool to select plants for resistance.

It is noteworthy that this wall coating can occur on the walls of infected vessels that are in immediate contact with another vessel and there are no intervening parenchyma cells to synthesize and extrude them. The lipoidal nature of these coating substances presumably allows them to creep along the surfaces of membranes and walls as they interface with the aqueous environment of the vessel lumen. Since oil spreads rapidly and extensively on water surfaces because of surface tension between the two, the creep could extend rapidly and for some distance beyond the source.

Within space-0, there occurs also an accumulation of various secondary metabolites, including the phytoalexins, and enzymes that are variously injurious and toxic to different pathogens and potential pathogens. These are noted at E and F in Fig. 2. β-1,3-glucanases and chitinases accumulate in plants treated with stress agents or infected with various fungi, including plants infected with *Fusarium* and *Verticillium* as well as with bacteria and viruses (see Krebs and Grumet, 1993). These enzymes degrade hyphal walls. Ferraris *et al.* (1987) found that infection with *F.o.l.* caused several to tenfold increases in chitinase, β-1,3-glucanase, glucosidase and *N*-acetyglucosaminidase activity in susceptible and resistant tomato cultivars from 5 to 19 days after inoculation. These enzymes were apparently of host origin. The greater activities were consistently found in the susceptible cultivar. Ferrafis *et al.* (1987), therefore, discounted the role of the enzymes in resistance and associated them with symptom induction. Unfortunately, the activities of these enzymes were not measured during the first few days after inoculation. It seems likely that if they are a part of the resistance complex, the enzymes would have been produced most strongly during the first 1–3 days after inoculation, after which, with the infection contained and isolated, their production would be cut off.

The continuous increase in activities found within the susceptible host could result from the failure of containment and the continued colonization of ever larger amounts of vascular tissue. This line of reasoning is supported by the findings of Benhamou *et al.* (1990) who found accumulation of chitinase in incompatible tomato − *F.o.l.* or *F.o. radicis-lycopersici* interactions to be faster than that in a compatible one. Using a glucanase or chitinase antiserum complexed with gold, the enzyme activities were found to be correlated with the distribution of the pathogen where host cell walls were in close contact with fungal cells. The early appearance of the β-1,3-glucanase was associated with the rapid localization of the infections in resistant interactions. The degree to which these enzymes are lethal or directly injurious to pathogens is still not clear, but they could be highly effective in cleaving oligosaccharides from fungal walls. These oligosaccharides could, in turn, elicit defense responses (Krebs and Grumet, 1993). This topic is pursued in more detail below.

The large and diverse family of phenolic metabolites, which are also involved in the processes of lignification and suberization, have been well reviewed with broad perspective by Mace (1989), and Nicholson and Hammerschmidt (1992). The great significance of these compounds lies in the fact that, once oxidized, they are highly reactive and combine chemically with a variety of complex carbohydrate, protein and lipid compounds, and with each other, to form macropolymers. In the process, enzymes are inactivated (Beckman, 1987) and the order of molecular complexity becomes so great that substrate polymers become almost immune to degradation and structures are made impermeable.

Jordan et al. (1988), in their study of vessel-coating and occluding masses in Fusarium-infected celery and celeriac, showed that the masses contained carbohydrate and phenolic substances and, in some cases, lipids. Significantly, they also contained granular centers of peroxidase activity. Thus all the elements were brought together on vessel walls and in vessel lumina by which a process of complex polymerization could, and apparently did, occur. Niemann et al. (1991b) analyzed the free and cell-wall bound phenolics in healthy and fungal-infected carnation. They found, by pyrolysis-mass spectroscopy, phenolic compounds bound, as esters, onto wall polysaccharides and significant increases in vanillic and ferulic acid content following infection. They suggested that these compounds might also be linked to cell walls, along with fatty acids, to suberize infected tissues.

Included among constitutive, secondary metabolites are antibiotic compounds such as tomatine and monomeric polyphenols such as caffeic and chlorogenic acids. The latter compounds have limited toxicity to wilt fungi, but their oxidation products are antibiotic (Mace, 1989). Thus, the decompartmentation and oxidation of these compounds following infection would enhance the plant's system of defense.

Among the complex of secondary metabolites that are produced in response to vascular infections are numerous other phytoalexins that accumulate within space-0 and serve, variously, to inhibit or kill many of the microbes that enter the xylem elements. Many of these are sesquiterpenoids which increase dramatically in resistant cultivars between 24 and 72 h after inoculation, but much more slowly in susceptible cultivars (see Beckman, 1987). This accumulation of the fungal inhibitors in the resistant cultivars corresponds to the rapid decline in fungal population in a resistant cultivar during the same time frame of 1 to 3 days (Shi et al., 1993). Among the best-studied systems of interaction is that of cotton infected with Verticillium (Mace, 1989) or Fusarium (Zhang et al., 1993). Here the phytoalexins are primarily desoxyhemigossypol and desoxymethoxyhemigossypol. Only desoxyhemigossypol was sufficiently soluble in water at the pH of infected stems to account for the death of V. dahliae conidia and mycelia at most sites in the stele of a resistant cotton cultivar (Mace et al., 1985). A complex of compounds including mansonones C, E, and F is produced in elm. In alfalfa, the phytoalexins are pterocarpans (Mace, 1989), and in tomato, rishitin (Elgersma and Liem, 1989).

Niemann et al. (1991a), in their studies on four cultivars of carnation, determined the secondary metabolites produced in xylem tissues following inoculation with races 1 or 2 of F.o. f sp. dianthi or Phialophora cinerescens. Samples taken at 4 and 9 days after inoculation showed accumulations of phenolic dianthramides that, based on statistical analyses, were determined partly by cultivar and partly by the invading pathogen. These accumulations were not related to the levels of symptoms that developed at 24 or 62 days after inoculation. It would be interesting, and probably pertinent to determine the levels of dianthramides that accumulated at 24–48 h in space-0, before any secondary distribution and extended elicitation of response occurred. A clearer picture of the role of these compounds might then be obtained. The synthesis of many of these compounds has been associated with the induction of enzymes of the shikimic acid pathway, notably phenylalanine ammonia lyase (Bernards and Ellis, 1991; Mace, 1989; Nicholson and Hammerschmidt, 1992). Clearly, these secondary metabolites must be added to the array of host defense factors in Fig. 2 and are noted at E.

The result of this array of recognition and response processes is that, for the most part, any parasitic organism that enters space-0 will be securely confined in a lateral direction and will be subjected to an environment that becomes biologically hostile to it. But what of the vertical direction?

3. Defining defense responses in space-1
Now that the infected vessel has been sealed off laterally, the problem that the plant faces in the vertical direction is the large open channel of the vessel above the trapping site in space-1 (Fig. 2). Somehow, this channel, and the transpiration stream that runs through it, must also be sealed off. Furthermore, this must be done before secondary spores are matured (48–72 h) and can be carried upward to the next trapping site some centimeters or meters away. Within 24 h after inoculation in a resistant interaction, however, the transpiration stream has been cut off by vessel-occluding gels (see Beckman, 1987; Charchar and Kraft, 1989). The gels arise by a process of swelling of end-wall and pit membranes and by *de novo* synthesis of primary cell-wall-like materials (Vander Molen *et al.*, 1986). Their production is mediated by ethylene (Vander Molen *et al.*, 1983). They are extruded by contact cells through pits (Moreau *et al.*, 1978; Shi *et al.*, 1992) in various resistant plant-parasite interactions. The gel-gum response is noted at G in Fig. 2.

This occlusion of space-1 by gels and gums is followed in many plants by an outgrowth of contact parenchyma cells through the pits and into the vessel lumina (Vander Molen *et al.*, 1987) where they balloon out, sometimes within the gels, to form tyloses. Where these tyloses come into contact, they fuse to form a tissue, thus walling off the affected vessels (see H in Fig. 2).

Numerous plants, including elm, oak, banana, tomato and squash, produce tyloses, but others, such as alfalfa, carnation, cotton, pea and radish do not (Beckman, 1987; Shi *et al.*, 1992). Chattaway concluded that the size of pit apertures determined whether tyloses or gums were produced in heartwood formation in trees (see Beckman, 1987). Recent results in our laboratory (unpublished) confirm this conclusion (Table I). Banana, melon and tomato, which produce both tyloses and gels, had oblong pits with an average aperture

TABLE I

Pit dimensions and calculated pit areas in the secondary xylem of tylose-producing banana, tomato, and melon plants, and in non-tylose-producing cotton, radish, tobacco, and pea plants

Plant Species	Pit dimensions		Pit area		Number of pits counted
	Range (μm^2)	Average (μm^2)	Range (μm^2)	Average (μm^2)	
Musa acuminata banana	4.0–25.0	4.1 × 13 (oblong)	15–100	53	200+
Lycopersicon esculentum tomato	1.6–17.0	3.8 × 10.3 (oblong)	6.4–102	39	118
Cucumis melo melon	0.7–10.0	1.5 × 7.0 (oblong)	2.1–100	49	83
Gossypium hirsutum cotton	0.2–2.0	0.86 (circular)	0.04–4.0	0.53	126
Raphanus sativus radish	0.1–3.5	0.77 (circular)	0.01–12.5	0.41	67
Nicotiana tobacum tobacco	0.1–1.9	0.68 (circular)	0.01–3.6	0.46	78
Pisum sativum pea	0.13–2.0	0.19 (circular)	0.02–4.0	0.031	40

size of 53, 49 and 39 μm^2, respectively. Cotton, tobacco, radish and pea, which do not produce tyloses, had circular, simple or heavily bordered pits with apertures of only 0.5, 0.46, 0.41, and 0.03 μm^2, respectively. Presumably, plants with pits so small as to prevent the passage of organelles necessary for cell growth cannot produce tyloses. The extruded materials may take on the appearance of tyloses by forming globular structures. These can only be distinguished with certainty from tyloses by TEM (Shi *et al.*, 1992).

Here we must modify our model (at H in Fig. 2) to depict tylose formation as an optional means for sealing off vessels in space-1. Gels are produced in any case. Hypertrophy of vascular parenchyma cells can still be an important part of the defense process, however, even in plants that do not produce tyloses. Primary vascular elements, having simple annular or spiral thickenings that are easily distorted, are often crushed by the growth of contact parenchyma cells following infection, particularly in resistant cultivars (Pennypacker and Leath, 1993). Such tissues were often found to test positively for suberin. Thus the gels, as well as tylose walls, finally become infused and polymerized with lignifying or suberizing agents to form the commonly reported gums associated with vascular infections. This is especially evident in wilt-resistant plants at the interface between space-0 and space-1. Suberization must be added to the host factors that confine infections in our model (E in Fig. 2).

In the final analysis, plants that are resistant to wilt diseases show a strong capacity to wall off space-1 above infected vessels with gels and tyloses or hyperplastic growth and, for some distance within the interface between space-0 and space-1, to lignify or suberize these structures. These responses are comparable to those in periderm formation and represent a basic mechanism of defense and healing common to all vascular plants.

Among the many secondary metabolites in plants are phenolics that are synthesized and stored in specialized cells during normal differentiation and maintained in a reduced state for extended periods by compartmentation within the cells. When the cells are subjected to stress, including infection, the stored phenolics are decompartmented, and, in cotton at least, encounter polyphenol oxidase that is associated with the thylakoids in plastids and peroxidase that is associated with cell walls (Beckman, 1987). Such compounds include the flavanoids catechin and gallocatechin in cotton, and dopamine in banana (Mace, 1989). These substances have the capacity, when oxidized, to polymerize with various macromolecules and with each other to form macromolecules of great complexity and resistance to degradation. Furthermore, the oxidation of such phenolics has been shown to mediate the conversion of tryptophan into indole-acetic acid, and indole-acetic acid to promote tylose formation (see Beckman, 1987). Thus, it appears that these commonly occurring phenolic-storing cells can serve a two-fold function, first by lignifying defense structures in space-0 and, second, by signaling for a growth response at a distance in space-1 that will seal off and lock up that avenue of advance. This hormonal response is noted at I in Fig. 2.

Thus far, we have dealt with the processes in space-0 and space-1 by which the host defends itself successfully. Section V will deal with those specific actions of a successful pathogen (on the right side of Fig. 2) that enable it to colonize the host and cause disease. First, however, it is useful to take a broader look at the nature of this colonization in terms of time and space within the host.

III.   THE RELATIONSHIP BETWEEN VASCULAR COLONIZATION AND FOLIAR SYMPTOMS

The localization of infections within vascular tissues has been well established as a primary mechanism of defense against vascular parasites in many plant species (see Beckman, 1987), particularly in plants showing high levels of resistance such as that afforded by the single-dominant I-1 gene in tomato against race 1 of *F.o.l.* It is not known, however, whether localization represents the sole or major mechanism of resistance, especially where polygenic resistance, sometimes termed intermediate resistance, is expressed. An altern-ative mechanism would be of considerable importance and scientific interest should it be clearly established. So far it has not.

Tolerance, by definition, is expressed when a plant, though extensively invaded, shows no or only slight to moderate symptoms and produces a reasonable crop. Gao *et al.* (1995a), compared the extent of vascular coloniza-tion and symptom expression in eight tomato genotypes: Bonny Best (highly susceptible), Marglobe and Improved Pearson (moderate polygenic resistance or "tolerance" to race 1), Improved Pearson VF-11 and Manapal (I-1 gene resistance to race 1), Rutgers (possibly containing the I-1 gene), Walker (I-1 and I-2 resistance to races 1 and 2), and 13R-1 (I-1, I-2 and I-3 genes for resistance to races 1, 2 and 3 of *F.o.l.*). These eight genotypes were inoculated with races 1, 2 or 3 of *F.o.l.* or with race 1 of *F.o. pisi* or *F.o. cubense* that are pathogenic to pea and banana, respectively. The degrees of symptoms expressed were rated at 28 days after inoculation, at which time the extent of colonization was also determined by aseptically cutting 3-4 mm thick stem sections at specified heights up the stem, fixing one end of the segments to microscope slides with lanolin, and incubating them for 24 h in moist chambers. A thin, free-hand sec-tion was then removed from the upper end of each segment, stained with cotton blue in lactophenol, and the number of infected vessels (from which hyphal tufts emerged) was counted by use of a light microscope. In all cases the degree of symptoms expressed was closely related to the extent of colonization of the host. A regression analysis comparing the uppermost extent of vascular colonization with the degree of symptom expression revealed a high degree of correlation, with an R value of 0.86 for 29 IP plants and an R value of 0.80 for 46 Marglobe plants ($P = 0.001$ for combined data). Therefore the degree of colonization appeared to be a major factor, if not the determining factor, for symptom expression, whether in a single-dominant-gene or polygenic interaction.

A colonization quotient (Q_{col}) was then calculated for each plant by dividing the number of infected vessels in a given sampling site (cross-section) by the number of infected vessels in the sampling site immediately below. Finally, an average quotient was calculated for each plant and cultivar/pathogen interac-tion (Table II). Similar calculations were made using the number of infected vessels at a given site at a given sampling date divided by the number at the preceding sampling date. Thus, the Q_{col} provided a numerical expression of the degree to which colonization was expanding (> 1), remaining relatively constant (≈ 1), or contracting (< 1) on the basis of height within the plant, or over time.

When the Q_{col} values were calculated in terms of progression up the stem, the Q_{col} for non-host interactions with *F.o. pisi* (Table II) and *F.o. cubense* (not shown) for all tomato cultivars was 0.00-0.01. Thus, there was essentially no advance of the vascular infections in non-host plants beyond the extent of initial uptake. This was equally true with the most susceptible cultivar, Bonny Best as with 13R-1 with its I-1, I-2 and I-3 genes for resistance to *F.o.l.* A clear demarcation (indicated by the bold line in Table II) was apparent between

TABLE II

Colonization quotients (Q_{col}) *that resulted when several tomato cultivars, having polygenes and I-1, I-2, and I-3 genes for resistance were exposed to* Fusarium oxysporum *f. sp.* pisi *or* Fusarium oxysporum *f. so.* lycopersici *(F.o.l.) race 1, 2 or 3. The known genes for resistance of each tomato cultivar are indicated in the second column; "I" indicates whether an appropriate single dominant "immunity" gene for resistance to a given race of the pathogen was present.* Q_{col} *values greater than 1 indicate that colonization is expanding, whereas those less than 1 indicate that colonization is contracting as the infection extends upward from sampling site to sampling site. The bold line separates apparent polygenic-type interactions (above) from single dominant-gene resistant interactions (below). Very low quotients indicate little or no progression of the infection, i.e., essentially complete localization.*

Cultivar	Genotype	Colonization quotient (Q_{col})*			
		F. o. pisi, race 1	F. o. l, race 1	F. o. l, race 2	F. o. l, race 3
Bonny Best	Highly susceptible	0.01	1.65	1.35	1.53
Marglobe	Polygenic	0.01	0.85–1.36	1.17–1.39	1.13
Improved Pearson	Polygenic	0.01	0.96–1.22	0.93–1.3	1.31
I.P. VF-11	I-1	0.00	0.13	0.31–1.08	1.04–1.40
Rutgers	I-1?	0.01	0.23	1.06–1.93	0.83–1.51
Manapal	I-1	0.00	0.36	1.22	1.39
Walter	I-1, 2	0.00	0.24	0.01	1.32
13R-1	I-1, 2, 3	0.00	0.22	0.04	0.22

* The colonization quotients (Q_{col}) were derived by dividing the number of infected vessels in a given cross-section of hypocotyl or stem by the number of infected vessels in the sample immediately below it.

cultivars that had (or apparently had in the case of Rutgers) an I-gene for resistance and cultivars that had polygenic resistance. Where the appropriate I-gene was present, the Q_{col} ranged from 0.01 to 0.36, that is, the plants showed a strongly contracting pattern of colonization. No symptoms were expressed in these plants. Where an appropriate I-gene was not present, colonization was considerably greater, but often variable, ranging from a continuing colonization ($Q_{col} \approx 1.0$) to a rapidly expanding pattern of colonization ($Q_{col} = 1.9$). The cumulative effect of such expansion in terms of time and space is given in Fig. 5.

A considerable range of Q_{col} values was common among the polygenic interactions where no appropriate I-gene was superimposed (e.g., interactions above the bold line in Table II). For the Marglobe/race 1 interaction the Q_{col}

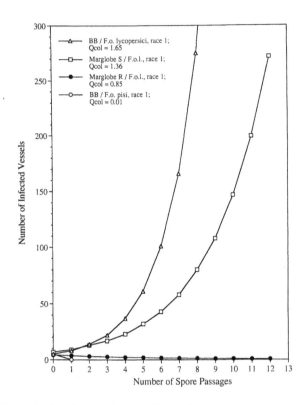

Fig. 5. The calculated cumulative effect of successive spore passages on the number of infected vessels based on various colonization quotients (Q_{col}) that were derived from four different host–parasite interactions and that represent a highly susceptible interaction (Bonny Best/*F.o.l.*, race 1), a moderately susceptible interaction (Marglobe S/*F.o.l.*, race 1), a moderately resistant interaction (Marglobe R/*F.o.l.*, race 1), and a non-host interaction (Bonny Best/*F.o. pisi.*, race 1). An initial number of five infected vessels was assumed in each case.

for individual plants ranged from 0.85 to 1.36, which represents a considerable difference in the pattern of colonization (Fig. 5). The Q_{col} in the IP/race 1 interaction ranged from 0.96 to 1.22, that for Rutgers/race 2 ranged from 1.06 to 1.93, and that for VF-11/race 2 from 0.31 to 1.08. This variation in Q_{col} was again closely correlated with symptom expression.

IV.   COLONIZATION IN SINGLE GENE AND POLYGENE RESISTANT PLANTS

The results of the experiments described in Table II make it clear that all the tomato cultivars tested, including the most susceptible Bonny Best, have the capacity to localize vascular infections in equal measure (Q_{col} values of 0.00–0.01 when infected with *F.o. cubense* or *F.o. pisi*). The differences appear to depend on whether the processes for localization in the plant are called into play or not, and how soon. Here, it seems, is where the control afforded by various I-genes and polygenes is played out. The results suggest that their function lies in the processes of recognition and transduction that, in turn, mobilize the processes of defense. Is this true also of "tolerance"?

There has been considerable conjecture, but little evidence, that tolerance might be distinct from resistance attributable to the processes of localization as described above. By definition (Anon., 1973), tolerant plants should be relatively free of symptoms, in spite of considerable colonization by the pathogen. That is, the host tolerates the presence of the pathogen. What is needed to establish tolerance, then, are statistically reliable data on the degree of colonization in relation to the degree of symptom expression. The nature of tolerance has been investigated by Gao *et al.* (1995b). The polygenic cultivar Marglobe was selected for this study because it was the most commonly grown "tolerant" cultivar for several decades before I-gene resistance was extensively employed (Walker, 1971). They grew sets of 20 seedling tomato plants of the cultivar Marglobe and removed and propagated axillary shoots as they enlarged to create families of genotypically identical plants. The parent plants were inoculated with race 1 of *F.o.l.,* and in subsequently repeated experiments, the vegetative propagants were inoculated with races 1, 2, and 3 of *F.o.l.*

Four weeks after inoculation, three parent plants were selected as being highly susceptible to race 1. Their vegetative progeny were rated an average of 3.8 for foliar symptoms (4.0 = 100% symptoms) with a maximum colonization height of 99%. The vegetative progeny of three Marglobe parents selected as resistant showed a foliar disease index of 0.9 (less than 25% foliar symptoms) and a maximum colonization height of 59%. In this experiment the vascular bundles of leaf petioles were also examined for infection. It was then apparent that the plants could sustain considerable vascular invasion of stems, where alternate pathways for transpiration were available, without expressing foliar symptoms. They displayed foliar symptoms only when vascular bundles of the petioles, especially the central bundles, were infected.

When average colonization quotients were calculated on the basis of initial and four weekly samplings, the weekly colonization quotient for Bonny Best (used as a check) was 3.7, that for a selected "susceptible" clone of Marglobe was 3.4 and that for a selected "resistant" clone of Marglobe was 1.9. Thus, the weekly rate of expansion of vascular colonization was considerable in the highly susceptible clone. Colonization was much slower in the more resistant clone, and the time at which petiolar infection led to symptom expression was considerably delayed. These resistant plants could thus be described as "slow wilting" plants.

The relative rate of colonization is determined, in turn, by the relative chance (as expressed in the Q_{col}) that an infection in a given vessel will advance or be contained. We therefore concluded that the term "tolerance" can be appropriately used to describe a cultivar that is polygenically diverse with respect to resistance. The term cannot be appropriately used to describe an individual plant within that cultivar. Each plant would better be described as more resistant or less resistant. The use of the colonization quotient would be the most accurate means (to date) for indicating that relative resistance.

V. HOW THE PATHOGEN COLONIZES THE HOST

So far, we have dealt with those responses of a plant that serve to wall off and isolate a potential pathogen. In this section we will examine the interaction as it appears when the pathogen successfully evades or disrupts these defenses to achieve systemic colonization of the vascular system. In reality the difference appears to be quite simple.

As we have seen in the preceding section, the differences between highly or moderately resistant interactions on the one hand, and highly or moderately susceptible interactions on the other, result from quantitative differences in the Q_{col} that describes, numerically, the rate at which vascular infections will contract or expand in number in a given plant, and at what relative rate. This, in turn is dependent upon the rates and degrees to which host defense responses are turned on, versus the rates and degrees to which the invasive weapons of the pathogen are turned on (left vs right sides of Fig. 2).

In the earlier model (Beckman, 1987) the capacities of the pathogen to invade were given as: (1) the rapid cycling of germinating spores that penetrate vessel endings and reproduce successive generations of spores; (2) the production of hydrolytic enzymes that could degrade vessel walls and gels; and (3) the production of inhibitors of respiration, callose deposition and tylose formation. The rapid cycling of spores gives the pathogen enormous mobility, if the pathogen can somehow outrun or delay or disrupt the localizing responses.

It has already been shown that different genotypes of the pathogen interact to signal their presence to different degrees in different genotypes of the host. The defense responses, all of them, are clearly slower or less intense in susceptible interactions (designated as "diminished" in Fig. 2). They are either not as strongly elicited or they are inhibited by a virulent pathogen. A tylose inhibitor produced by the pathogen had been postulated in the earlier model because it had been shown that tylose formation, though initiated equally well in resistant and susceptible lines of tomato, was inhibited after day 2 in the susceptible cultivar (see Beckman, 1987). However, Harrison and Beckman (1987) showed that the inhibitor was rishitin, a host product. Obviously something was amiss. They proposed, therefore, that the rapid degradation of vascular gels by fungal hydrolytic enzymes caused the vessel-occluding gels to shear under transpirational pull. This would, in turn, cause rishitin, which plays an important role in locking up the defense structures in space-0, to be displaced upward throughout space-1 where it would lignify the walls of developing tyloses, inhibit their enlargement, and thus disrupt the entire walling-off process.

The pathogen factors for successful colonization shown in Fig. 2 have thus been revised to: (1) minimizing recognition and the resulting host responses (by mimicry?); (2) the production of hydrolytic enzymes; (3) the rapid cycling of successive generations of spores; and possibly (4) the capacity to detoxify phytoalexins. Of these pathogenic factors, the minimizing of recognition and host response is clearly suggested by the studies of Beckman *et al.* (1989, 1991). The common antigen studies of DeVay and colleagues (DeVay, 1989) suggest that the minimizing of recognition could be achieved by the pathogen through molecular mimicry. This hypothesis is attractive in that the possible variations in proteins, glycoproteins, lipoproteins, and lipopolysaccharides are endless and could easily account for varying degrees of host specificity. Thus "mimicry (?)" has been added to the revised model (Fig. 2) as a likely determining factor in pathogenesis.

The second pathogen factor, the production of hydrolytic enzymes, has been studied extensively and has also been reviewed in terms of colonization of the vascular systems of plants (Durrands and Cooper, 1988). As discussed earlier, except for the absence of crystalline cellulose, the occluding gels of space-1 contain polymers comparable to those of primary cell walls, i.e., pectins, pectates, and various hemicellulose constituents (Vander Molen *et al.*, 1986). The primary enzymes produced by the pathogens within vascular tissues are endo-forms of pectolytic and hemicellulolytic enzymes. The activity of such enzymes causes a rapid loss of macromolecular integrity of the gels because the molecular length of their constituents and their resistance to shear is greatly reduced. Durrands and Cooper (1988) demonstrated that the loss of any single enzyme by a pathogenic isolate of *V a-a* reduced its virulence, that is the rate at which it would produce symptoms. Clearly, the loss of one of several hydrolytic enzymes could reduce the rate of macromoleculer cleavage in the gels and delay the loss of shear resistance. The rate of colonization, the Q_{col} would thus be lowered. They showed, furthermore, that a secretory mutant of the pathogen, one that was unable to secrete any of these enzymes, failed to cause disease. A conflicting factor in this assessment is that they were unable to detect differences in host colonization between the wild type and the various enzyme-deficient mutants. Only the secretory mutant, which failed to excrete any of these enzymes, failed to colonize the host. A more critical analysis of colonization by determining Q_{col} values using hyphal tuft counts for viable infections, might afford a more accurate assessment of this question.

These lines of evidence suggest the following scenario for a susceptible interaction, as depicted at the right-hand side of Fig. 2. First, the host responses are delayed and diminished by mimicry or comparable strategy by the pathogen. Thus the host is less able to isolate the fungus from its nutrient base in contact parenchyma cells, fungal inhibition by secondary metabolites is diminished, gels are more slowly induced and the gels that are produced are only slowly lignified and suberized. The result is that fungal hydrolytic enzymes are much more actively induced in a susceptible than in a resistant interaction, the gel polymers that are produced are more actively cleaved and shear under transpirational pull, and the entire interaction of space-0 moves upward into and through space-1. That is, the interaction is displaced upward.

It is clear that the success or failure of the plant to localize infections, on the one hand, or the capacity of an adapted vascular parasite to colonize the vascular system of a plant, on the other, is in a delicate balance and it is so at every site of entrapment in every infected vessel of a plant. The Q_{col} represents the numerical chance (likelihood) that each infection at 1 or 10 or 100 successive sites will escape or be contained. Thus, in I-gene and polygenically resistant plants the factors of recognition and transduction become translated into responses that range from rapid, consistent containment of the pathogen to a halting or surging colonization of the vascular system by the pathogen, and, finally, a correlated rate and extent of plant dysfunction when the vascular systems of petioles and leaves are invaded.

VI. MOLECULAR ASPECTS OF WILT DISEASE

The structural, biochemical and Mendelian genetic observations described thus far inevitably lead to questions regarding the types of molecular interactions, both biochemical and genetic, that occur between a vascular parasite and its host. In any molecular model of disease resistance, the plant response will involve several steps: first, the production of a signal indicating the arrival of a potential pathogen; second, the perception of this signal and finally, signal transduction leading to a physiological response. Fundamental aspects that eventually must be included in a model of wilt disease resistance (or its failure) include the molecular identity of the signals passed between host and parasite, the mechanisms by which such signals are transduced and a more complete biochemical characterization of responses in both space-0 and space-1. In contrast to our understanding of the classical genetics of wilt resistance genes, the molecular role(s) that single dominant resistance genes and their products play in these steps is just beginning to be understood. The basis of polygenic resistance remains largely unaddressed. As progress in understanding the molecular aspects of wilt is made, part of our challenge will be to incorporate these findings within the framework provided by more classical approaches into a model of the disease process.

Although our understanding of the molecular biology of wilt disease resistance is less advanced than our structural and physiological understanding, the following elements might be included in a working model. A pathogen entering the vascular column of space-0 rapidly makes intimate molecular contact with the walls of vascular parenchyma cells of a host. When host and potential pathogen meet, many types of signals are exchanged. The first living host cells that receive these signals are likely to be paravascular parenchyma cells ("contact cells"). Because disease resistance is the rule and susceptibility the exception, it seems that many types of signals or perturbations are capable of activating signal transduction networks that in turn activate latent and transcriptionally dependent responses in a timely manner. Some of these signals may be received by specific receptors located at the plasma membrane of the host, whereas others may interact with plant cells in a less specific way that is not dependent on a specific receptor. Although genes for specific receptors involved in recognition of potential fungal vascular wilt pathogens have not yet been cloned, it is possible that they code for some of the single dominant I-genes that are well known from classical genetics (for example, see Gabriel et al., 1988; Keen, 1990). Other I-genes are likely to code not for membrane-bound signal receptors, but rather for proteins involved in signal transduction pathways (Martin et al., 1993). In either case, triggering of signal transduction pathways leads to activation of latent defense processes such as callose synthesis and deposition, and subsequently alters patterns of gene transcription and translation. Concurrent with these steps, the affected plant cell(s) send chemical signals to cells upstream into space-1 of the vascular system. Here, a totally different set of responses occurs. It is likely that both susceptible and resistant plants have the cellular machinery needed to resist wilt pathogens; it is their timely activation and the resulting differences in Q_{col} that may ultimately determine the outcome of the host–pathogen interaction (Beckman et al., 1989). Successful pathogens may produce fewer signals or otherwise reduce the level of perturbation, perhaps by avoiding triggering a response (molecular mimicry) or by active inhibition of the signal transduction or response processes (Fig. 2).

A. SIGNALS INVOLVED IN HOST/PATHOGEN INTERACTIONS IN SPACE-0

In general, active responses to a vascular pathogen begin when signals generated by the presence of the pathogen are recognized by the host. Signals capable of initiating resistance responses of the host are numerous. Many are suggested to be diffusible chemical substances called elicitors. A wide variety of such diffusible elicitors, including poly-and oligosaccharides, glycoproteins and proteins, have been shown to induce some or all of a host's defensive physiological responses to pathogens (e.g., Hahn et al., 1993). Many of these responses have been studied in vitro by use of plant tissue cultures exposed to elicitors of varying purity. Although in vitro work is essential, variation of cell type and spatial location probably also influence the progress of vascular disease in whole plants. Thus, a formidable research challenge is to determine which signals are required for the successful response of an intact plant to a challenge by a vascular pathogen. The ability to manipulate host genes involved in the defense process will be useful in testing many hypotheses.

Oligosaccharide elicitors have received considerable attention as potential molecular signals. These carbohydrate fragments are attractive candidates as signaling molecules because they can be produced by enzymatic degradation of either the fungal cell wall or the plant cell wall (Ryan and Farmer, 1991). Many plant cells constitutively produce low levels of enzymes capable of degrading common fungal cell wall polymers such as chitin or chitosan. The enzymatic liberation of fragments of these walls may serve as useful fingerprints that alert the plant to the presence of a wide range of potential fungal pathogens. Conversely, the action of fungal enzymes essential for penetration of the host cell walls may liberate oligosaccharide fragments from these walls (Ryan and Farmer, 1991). In either case, the oligosaccharide signals produced are free to diffuse from the site of production through the apoplast and rapidly arrive at the plasma membrane of a living host cell (contact cell) where they serve as a signal capable of inducing a response.

While the idea of a diffusible elicitor of some type is widely accepted, structural observations of host/pathogen interactions in vascular wilt disease suggest that non-diffusible signals might also play a role in signaling. As noted earlier, when fungal propagules are introduced into the vascular column of a host, most accumulate rapidly below the end walls of vessels. Some become attached in other localized regions of the xylem, especially near pit fields between adjacent vessels (Beckman et al., 1989). Electron microscopy shows that the wall of the pit becomes concave and conforms closely to that of the fungal spore indicating a strong molecular interaction (Beckman et al., 1989). Similar adhesion to pit sites of contact cells occurs preceding the early stages of a host response. Cytoplasmic rearrangement and deposition of callose often appear to be localized in cells at sites that are directly opposite fungal pathogens (Beckman et al., 1991). This response may be localized because host cells closest to the fungal pathogen are the first to receive diffusible elicitors. Alternatively (or in addition), larger-scale interactions between the cell walls of host and pathogen could be involved. Contact and interaction between the surfaces could result in molecular stresses or dislocations — "force fields" — that are able to act as signals to the host. Signals generated by membrane deformation are thought to be involved in plant thigmotropic responses; pathogen attack may also cause membrane stretching that initiates signaling (Weiler, 1993). A mechanical component to signaling would be consistent with observations that some non-host pathogens fail to induce a defense response from host tissue if they are separated by dialysis membrane, even though soluble components would be freely diffusible (Hadwiger, 1991).

How might the close physical binding of a vascular pathogen to a host cell wall produce a detectable signal? One possible mechanism of such molecular binding might involve molecules similar to vitronectin, a glycoprotein that occurs in the extracellular matrix of animal cells, and is involved with adhesion and cell signaling. Vitronectin-like molecules have also been identified in walls of plant species, such as tomato (Sanders *et al.*, 1991). Vitronectin has recently been implicated in the attachment process of *Agrobacterium tumefaciens* to suspension cultured carrot cells (Wagner and Matthysse, 1992). Wagner and Matthysse (1992) have speculated that, in analogy with the role of vitronectin in animal cells, apoplastic vitronectin may be linked with the cytoplasm of the plant cell by molecules such as integrin. Membrane bound integrins might then transduce a signal to the cytoplasm, thus signaling the arrival of potential pathogens within the vascular system of the host. Clearly, such a proposal cannot account for all signals (for example, some protoplasts can respond to elicitors) but it is useful to remember that a complex response in a living plant may depend on many types of signals produced by a pathogen. Indeed, the wide variety of chemical and abiotic elicitors capable of inducing defense responses, and the variable degree of plant response, suggest that many signals may stimulate a common signal transduction pathway that activates some minimal suite of physiological stress responses.

B. SIGNAL PERCEPTION IN HOST CELLS

Whatever the nature of the elicitor, the signal produced is useless until its potential information is conveyed to the cytoplasm of a living cell that is capable of physiological response. There are several ways that such information can be transmitted, but each begins when the elicitor interacts with the plasma membrane of a host. One commonly accepted hypothesis is that elicitors are recognized and bound by specific receptors found in the plasma membrane (e.g., Gabriel *et al.*, 1988; Kleen, 1990). Binding of an elicitor would trigger some change in the receptor protein leading to transduction of a signal to the cytoplasm. The existence of specific receptors is consistent with observations that the ability of some elicitors to stimulate defense responses can be altered by small chemical modifications to its size or shape (Hahn *et al.*, 1993).

The many elicitors that are capable of inducing defense responses suggests that other less specific types of signal perception can occur. Elicitors may be recognized by more general, non-specific mechanisms. Direct interaction of an elicitor with the host membrane itself could lead to a perturbation that may constitute signal perception. The direct entry of an elicitor into the cytoplasm of a host cell represents another mechanism that could signal the presence of a pathogen and initiate a defense cascade.

If specific plasma membrane-bound receptors form a functional part of the signal recognition process, they would provide a potential explanation for the race specificity exhibited by certain vascular pathogens (Gabriel *et al.*, 1988; Keen, 1990; Sutherland and Pegg, 1992). Resistant plant cultivars might possess a receptor that confers the ability to recognize a specific race of pathogen and invoke defense reactions in a timely fashion leading to a low value of Q_{col}.

Such receptors might be missing from susceptible cultivars, which could cause recognition and response to occur too slowly to prevent the spread of the pathogen, as indicated by a high Q_{col}. Although the concept of specific plasma membrane receptors is appealing, they have proven difficult to isolate. Several receptors for race-specific fungal elicitors have been reported (Dixon and Lamb, 1990), but specific receptors involved in fungal wilt disease have not yet been isolated. If specific receptors are shown to be important for recognition, their isolation and characterization clearly would be a significant advance. Many questions could then be addressed. Would all vascular parenchyma cells possess such receptors? What density of such receptors would be required to detect the low levels of elicitors that are produced in the early determinative stages of infection? Progress in biochemistry and molecular genetics may soon allow us to begin to include these in our model.

Despite its capacity to explain race specific resistance, the specific receptor model is not without problems. As noted by many authors, it is unlikely that each of thousands of potential pathogens can be recognized by its own distinct receptor simply because the number of receptors would be astronomical. Perhaps some common characteristics of fungal pathogens allow large groups to be identified, whereas only a limited number of pathovars have evolved to evade or suppress host recognition and response and thus become host specific pathogens. For example, the common fungal cell wall components chitin and chitosan (or oligosaccharides released from them) may interact with the plasma membrane in a manner that does not depend on specific "receptors". Fragments of chitosan have been shown to induce the accumulation of phytoalexins, proteinase inhibitors and callose in several plant species (Hahn, et al., 1993). The ability of such compounds to act as elicitors of callose synthesis was shown to vary directly with molecular weight and the structure of the polysaccharide (Kauss, 1990). Callose synthesis was found to increase as the degree of N-acetylation of chitosan decreased, suggesting that availability of amino groups might be important for binding. Kauss (1990) suggested that one mechanism by which chitosan elicitors interact with host cells is not via a specific receptor but, rather, by means of a non-specific interaction with the lipids of the plasma membrane. This interaction may be ionic, the positively charged amino groups of chitosan binding with the negatively charged phospholipids of the plasma membrane. As a result of this interaction, the membrane becomes "leaky", leading to an influx of ions (e.g., Ca^{2+}) and cell depolarization. Higher molecular weight chitosan oligomers would presumably be able to interact with greater numbers of lipids, thus leading to greater membrane perturbation and increased response. Signal reception would then acquire a quantitative character (as indicated by Q_{col} values). The suggestion that non-specific interactions may generate a signal is supported by experiments in which uncharged amphipathic compounds, such as digitonin or tomatin, that also alter membrane integrity, are capable of inducing callose synthesis as well (see Kauss, 1990).

Interactions between elicitors and host cells do not necessarily end at the host plasma membrane. Some studies have indicated that after elicitors meet the membrane, they may be internalized by endocytosis. This suggestion is based in part on observations by Hadwiger and coworkers (Hadwiger, 1991) that, following contact between the fungal pathogen *F. oxysporum* f. sp. *pisi* and pea, radioactively labeled chitosan oligomers could be detected in the nuclei of plant cells within 30 min. The mechanism of uptake is not understood, nor is it clear if plasma membrane receptors with a specific affinity for chitosan are required.

C. SIGNAL TRANSDUCTION AND CELLULAR RESPONSE

The signals that operate in the early stages of vascular infection are likely to be produced in small amounts. For this reason, once a signal (the "primary messenger") is perceived, it must be amplified. This amplification can be accomplished by a signal transduction cascade producing "second messengers" that mediate the intracellular response. Various second messengers may participate in a successful plant response to infection, including membrane depolarization, ion fluxes, phosphoinositides, jasmonates and others (Lamb *et al.*, 1989; Dixon and Lamb, 1990; Trewavas and Gilroy, 1991; Atkinson, 1993; Drøbak, 1993). Signal transduction pathways may interact, creating a signal transduction network rather than a simple linear chain of events. The differential activation of branches of these networks may account for the variable resistance of plants to different races of a vascular pathogen. In support of this idea, recent studies have suggested that the products of single dominant genes conferring race-specific resistance against a bacterial pathogen may be directly involved in signal transduction (Martin *et al.*, 1993). In the following paragraphs, selected aspects of signal transduction and second messengers that may play a role in resistance to vascular (and other) diseases are highlighted. No doubt others are critical to a successful defense response.

Following signal perception, one signal transduction event that can be induced is an alteration in ion flux at the plasma membrane. Such responses have been observed following treatment of plant cell cultures with fungal elicitors. The membrane depolarization that results can be quite rapid although it is often transient or oscillating (see Gilroy and Trewavas, 1990). One ion that has received special attention as a modulator of cell defense responses (as well as many other non-stress responses) is Ca^{2+}. The involvement of calcium ions as second messengers in both plant and animal cells is widely recognized. Calcium displays many properties useful for second messengers. It occurs in low concentration in the cell cytoplasm, but can readily be mobilized from intracellular storage sites (vacuoles or endoplasmic reticulum) or from the apoplast where it occurs in relatively high concentrations. It can be transported and therefore regulated by various channels and pumps and, equally important, can stimulate cell responses through interaction with calcium-binding proteins. It is likely that Ca^{2+} is involved in some of the early microscopically visible responses observed in plants colonized by a vascular pathogen, including cytoplasmic rearrangement (after 30–60 min) and callose deposition (4 h) (Beckman et al., 1991; Mueller et al., 1994). These responses may represent the activation of preformed (latent) defensive systems that can be mobilized without the need for gene transcription and translation. Callose synthesis in plants is known to be Ca^{2+} dependent and it has been suggested that its deposition could be the result of an elicitor-stimulated increase in cytosolic Ca^{2+} (Kauss, 1990). In tomato/*Fusarium* interactions, as well as in many other host/pathogen interactions, callose accumulates in regions of the host cytoplasm directly opposite a fungal spore (Beckman et al., 1989). Such a pattern may be the result of localized increases in cytoplasmic Ca^{2+} concentrations, perhaps due to localized opening of ion channels in the plasma membrane. Although Ca^{2+} has not been directly implicated in the cytoplasmic rearrangement observed in tomato, it may regulate motor proteins and thus organize host cytoplasmic machinery to block the progress of a potential pathogen. Calcium ions have the potential for modulating numerous other cellular responses, including gene activation. Some of these effects may be directed by the calcium binding protein calmodulin that acts as a calcium-dependent switch for various enzymes including protein kinases. Activation of protein kinases could amplify the Ca^{2+} signal by turning on enzymatic and genetic systems involved in the production of defensive compounds. Molecular evidence that the gene product of a single dominant resistance locus is a protein kinase participating in this signal transduction pathway has been reported (Martin et al., 1993). Although much research on the possible role of Ca^{2+} and other ion fluxes in plant defense responses remains to be done, it is likely that they will form a part of molecular model of vascular wilt response.

Another signal transduction pathway and group of second messengers that may play a role in transducing an extracellular signal and activating latent and transcriptionally based responses involves inositol phospholipids. In a typical cascade, signal perception leads to the activation of phospholipase C enzymes associated with the plasma membrane of the host. These activation steps may involve the participation of G-proteins (regulatory elements that may modulate phospholipase activity). The phospholipase cleaves the plasma membrane lipid phosphatidylinositol diphosphate producing inositol triphosphate (IP3) and diacylglycerol. Each of these products may, in turn, play a number of regulatory roles. For example, it is suggested that IP3 stimulates the release of Ca^{2+} from intracellular stores thus activating protein kinases and ultimately leading to a cellular response. Each step of this transduction pathway allows the potential regulation and specialization of host response to an invading pathogen.

In some cases, it may be possible that an elicitor can interact directly with host cytoplasmic components thus bypassing second messenger pathways. One example has been described by Hadwiger (1991). As previously noted, when peas are exposed to the non-host pathogen *Fusarium solani* f. sp. *phaseoli*, chitosan derived from the fungal cell walls rapidly accumulates within the host nucleus. Here, Hadwiger hypothesizes that chitosan may interact directly with host DNA, possibly binding short chromosomal loops. Binding, in turn, alters the DNA conformation leading to the activation of defensive genes ("slave genes"). Activation of slave genes can also be controlled by single dominant resistance genes ("master genes"). Possibly, some master genes can be directly activated by certain elicitors, although this has not been demonstrated.

Although studies on the signal transduction pathways involved in the interaction between host and vascular pathogen are of great interest, their biochemical analysis presents many difficulties. Our understanding of the biochemistry of this process, and elaboration of our model of the entire disease interaction, will no doubt be assisted by the ability to examine and manipulate the genes involved in defense. Some of these are described below.

D. MOLECULAR GENETICS OF RESISTANCE

Presently, our knowledge of resistance to vascular parasites is primarily limited to single dominant genes referred to as "immunity" genes (I-genes). Studies of the genetic basis for host/pathogen interactions have been extensively reviewed (Keen, 1990; Briggs and Johal, 1994). According to current understanding, the outcome of the encounter between a host and its potential pathogen is determined by the interaction of the products of host resistance genes and corresponding "avirulence" genes within a pathogen. Mutations or deletions of avirulence genes allow a pathogen to grow on a previously resistant host consistent with the classic gene-for-gene relationship described by Flor (1942).

Although several proposals have been put forward to account for the action of I-genes, the suggestion that these genes code for plasma membrane-bound proteins that are involved in signal recognition has received much consideration (Gabriel *et al.* 1988; Keen, 1990). This hypothesis is appealing, in part because it seems to fit well with the gene-for-gene relationship that is observed in single dominant gene resistance.

Although the hypothesis that single dominant resistance genes may code for specific membrane-bound signal receptors is appealing, recent studies on the molecular genetic basis of such resistance indicate that single dominant resistance products may be involved in the signal transduction pathway (Martin *et al.*, 1993). The interaction of *Lycopersicon esculentum* and the bacterial speck pathogen *Pseudomonas syringae* pv *tomato* is determined by the interaction of a single dominant gene in the host (the Pto locus) and the corresponding avirulence locus in the pathogen (avrPto). A putative *Pto* gene has been isolated from a resistant tomato cultivar using RFLP markers to screen a tomato genomic library. The identity of the cloned *Pto* gene was tested by transforming a susceptible near isoline. The transformants were found to be resistant to *P. syringae* pv *tomato* and this resistance was inherited in a Mendelian fashion.

Analysis of the *Pto* gene sequence and comparison of the deduced amino acid sequence with other known proteins has revealed several intriguing properties (Martin *et al.*, 1993). The protein appears to be hydrophilic and, although it may be membrane associated, it contains no obvious membrane spanning domains, thus indicating that it is unlikely to act as a transmembrane signal receptor at the plasma membrane. Equally intriguing, comparison of the *Pto* gene sequence with other proteins revealed similarity to the catalytic regions of serine–theonine kinases of plants and other organisms. Serine–threonine kinases are enzymes that can function in signal transduction by activating or deactivating enzymes or genes. A role in recognition is suggested by the fact that the Pto protein of tomato shows similarity to a serine-threonine kinase from *Brassica* that is believed to play a role in pollen incompatibility of stigma cells (Stein *et al.*, 1991).

The idea that single dominant resistance genes may code for proteins involved in signal transduction rather than recognition has several pertinent implications for understanding vascular wilts. Transfer of the *Pto* gene alone to a susceptible near isoline was able to confer a resistant phenotype. This implies either that a specific "signal receptor" is not required for the response, or that such receptors already exist even in susceptible plants. The mechanism of the I-genes involved with resistance to vascular disease remains unknown, although attempts to clone such genes are currently in progress (Segal *et al.*, 1992). As genes involved in resistance to vascular wilt pathogens are cloned and analyzed, it will be of great interest to see if they, too, code for proteins forming part of the signal transduction pathway, or if membrane bound signal receptors will be found.

Evidence has also been presented that Pto is one member of a larger family of related proteins in tomato. Using the *Pto* gene as a probe, Martin *et al.* (1993) were able to identify at least six other homologous genes in tomato genomic DNA from both *Pseudomonas*-resistant and -susceptible cultivars. It is not yet known if these homologous sequences represent serine–threonine kinases or if any of them function in resistance to other potential pathogens. *Pto* and several of its homologous sequences were found to be clustered together on a 400 kb fragment of DNA. The clustering of these genes is consistent with the suggestion that genetic rearrangements between gene family members may have occurred and that such rearrangements may be responsible for generating new alleles capable of conferring resistance against new pathogens (Martin *et al.*, 1993). Understanding the regulation and function of these genes can be useful not only for the direct transfer of such traits, but also potentially to modify or construct new resistance genes tailored to certain pathogens.

VII. INTEGRATING HOST RESISTANCE AND OTHER FORMS OF BIOLOGICAL CONTROL

This section considers the various strategies for controlling wilt disease in light of the foregoing understanding of how plants defend themselves against vascular parasites (or, fail to do so). We believe that when the physical or biological environment is manipulated the capacity of the plant to respond and defend itself is altered. The effects of cross-protection, induced resistance, suppressive vs. conducive soils and soil management, including soil solarization, will be examined from that point of view. It will then be shown whether our contention is consistent with the evidence at hand and benefits our understanding.

A. CROSS PROTECTION AND INDUCED RESISTANCE

It has long been known that the xylem elements of plants can be invaded to a limited degree by various fungi and bacteria (often through wounds) without causing serious disease symptoms, if any. It has also been shown that cultivars that are highly susceptible to a given form, race, or pathovar can be induced to a higher level of resistance (cross-protected) by preinoculation with a non-pathogenic or avirulent microbe, or one specific to another host (Matta, 1989). Such induced resistance has been shown to extend for some distance beyond the zone of infection, to require an optimum of one to a few days to become optimally functional (Biles and Martyn, 1989), and can persist for some time thereafter. Preinoculation apparently conditions the plant to respond more rapidly to subsequent infections by a pathogen (Matta, 1989), with all the previously described defense responses (callose deposition, gel and tylose formation, and secondary metabolism) being enhanced by the preinoculation (Matta, 1989).

This enhanced resistance can be induced by growth retarding chemicals (Cohen et al., 1987), by herbicides (Awadalla and El-Refai, 1992), by coating roots with antagonistic pseudomonads (Peer et al., 1991) and various avirulent and non-host pathogens (Biles and Martyn, 1989; Jorge et al., 1992). The elicitation of induced resistance is therefore non-specific. There is good evidence that ethylene may play a key role in promoting this enhanced resistance because ethylene is induced following many kinds of tissue disturbance (including physical injury, infection and chemical injury), and because ethylene treatment alone has been shown to promote many of these responses in plant tissues (Vander Molen et al., 1983; Chen et al., 1986; Matta, 1989). The use of this preinoculation technique may have limited application in the field, however, because of the transitory nature of the conditioning. A related approach of greater promise might be to alter the soil environment to shift and maintain a more favorable population of soil-borne organisms.

B. SUPPRESSIVE SOILS

Extensive studies of the nature of soil suppressiveness and efforts to manipulate soil to achieve suppressiveness have been reviewed by Alabouvette (1989), Lemanceau (1989), Louvet (1989), and Schneider (1982). When many soils are tested, a continuum from highly suppressive to highly conducive soils can be found (Alabouvette, 1989; Louvet, 1989). Rouxel *et al.* (1988) determined the levels of wilting on different soils with respect to time after inoculation and to inoculum levels. They went on to calculate plant survival probabilities as a means of obtaining numerical comparisons of the suppressiveness of various soils (see Louvet, 1989).

Many organisms contribute to suppressiveness in a given soil (Lemanceau *et al.*, 1988), but major contributors to Fusarium and Verticillium wilt suppressiveness in nature appear to be non-pathogenic forms of *F. oxysporum*, *Verticillium dahliae* and fluorescent pseudomonads (Louvet, 1989; Price and Sackston, 1989). The pathogen can be amply present in the soil and establish itself within the host, but fail to produce disease. A prominent difference between suppressive and conducive soils that could account for this failure is the ratio of pathogenic/non-pathogenic forms of *F. oxysporum*. In highly suppressive soils, the numbers of non-pathogenic forms are 10 times that in conducive soils (Louvet, 1989) and the competition for nutrients, especially carbohydrate and iron (Lemanceau 1989), seems to determine how these organisms function in the soil.

Louvet (1989) proposed that studies of interactions between *F. oxysporum* and root tips of the host could be pertinent to an understanding of soil suppressiveness. Huisman and Gerik (1989) demonstrated that germination and growth of resting chlamydospores of *F. oxysporum* occurred in response to the exudates released from a passing root tip. Thus, germination and hyphal growth have to be exceedingly rapid in order for successful root colonization to occur. Any biological activity in the soil that in some way affects this rapid response can determine the degree to which the pathogen gains entry into the host or, more significantly, the degree to which it gains entry into vascular elements.

Achieving vascular infections in nature may, however, be a two-step process. In banana, *Fusarium* infection of the vascular elements apparently occurs only through wounds and, in many other interactions, wounds greatly increase the incidence of disease (Beckman, 1987). We therefore propose a slightly more complex sequence of events. Wounds would greatly enhance vascular infection if the pathogen were present within the immediate vicinity of the wound and in the form of propagules that could be drawn, with soil and tissue fluids, into a severed vascular element. As Huisman and Gerik (1989) showed, certain microbial soil inhabitants are selectively induced to germinate and colonize epidermal and cortical tissues. If a rodent or grub feeds through such infected tissues into vascular elements, there would be direct entry of a large population of select forms of root inhabiting microbes into vascular elements. If, as in suppressive soils, that microbial mix is dominated by non-pathogenic organisms, especially non-pathogenic forms of *F. oxysporum*, *V. dahliae* or fluorescent pseudomonads, the resistance responses of the host would, presumably, be strongly triggered and the infections sealed off. Soil management, then, becomes a powerful tool in the management of wilt diseases.

C. SOIL SOLARIZATION

Soil solarization, a hydrothermal process by which soil is heated in the field for several weeks, has served to control many disease, insect, and weed problems, and to enhance crop production. Following rain or irrigation to bring the soil up to a high level of water holding capacity, the field is covered with transparent polyethylene sheeting to allow solar heating and enhance heat and moisture retention in the soil (Katan, 1989; DeVay, 1991). Solarization is practiced during warm summer months when solar radiation and ambient temperatures are high. During such periods, soil temperatures that are inhibitory and even lethal to many plant pathogens and pests are achieved to a considerable depth.

The effectiveness of soil solarization is based on the principle that most plant pathogens and pests are unable to grow at temperatures above 32°C. Thermotolerant and thermophilic microorganisms usually survive the soil solarization process (DeVay, 1991). Thus, the population of soil microflora is massively shifted toward non-pathogenic organisms. It would be interesting and significant to determine whether or not solarized soil takes on a suppressive character with respect to wilt and other forms of disease. In any event, the effects of soil solarization on disease suppression can persist for several years.

D. SOIL MANAGEMENT

That management of soil suppressiveness is feasible is suggested by numerous studies. Louvet and various colleagues (see Louvet, 1989) converted conducive soil into a suppressive condition by sterilizing and then adding as little as 1% of suppressive soil, together with a carbohydrate, to conducive soil to establish anew, suppressive microfloral population. Garibaldi *et al.* (1986) found that Fusarium wilt of carnation could be suppressed by adding saprophytic *F. oxysporum* or *F. solani* to steam disinfected soil or by dipping the roots of cuttings in a spore suspension of the antagonists before transplanting into infested soil. On a longer-term basis, field soils have been managed to reduce the levels of wilt disease in numerous crops, apparently by affecting the activities of soil microflora. This has been achieved by managing the soil pH by the addition of calcium, which, in turn, limits the availability of soil micronutrients (Engelhard *et al.*, 1989).

Schneider (1985) has demonstrated additional factors in soil that can influence the level of disease, e.g., the level and balance of K^+ and Cl^- ion availability. Plants adjust to changes in water availability by adjusting the osmoticum within their cells. This adjustment is made most easily and most efficiently by adjusting the levels of K^+ and Cl^- within their cells. When K^+ and Cl^- are inadequate or when Cl^- is deficient to the balance, organic acids are pumped out of root cells to adjust to the increased water availability. After a rain storm, this can cause a considerable increase in organic acids within the root rhizosphere, thus attracting and supporting a larger population of rhizoflora, including, in some circumstances, vascular pathogens.

A comprehensive program of soil management to minimize wilt diseases has been proposed by Alabouvette (1989) that includes increasing the general biomass of the soil with particular attention being given to increasing the population of specific microorganisms that have been shown to be suppressive. The ultimate objective of such a program would be to achieve stable suppressiveness.

VIII. THE EXPRESSIVE PHASE OF DISEASE DEVELOPMENT

Numerous studies show that foliar symptom expression, whether wilting, yellowing or necrosis, occurs after fungal colonization of twigs and leaf petioles and even portions of the leaf blade (see Beckman, 1987). An earlier hypothesis held that fungal "toxins" could operate at considerable distances beyond sites of colonization to cause loss of semipermeability of leaf cells, excessive loss of water and wilting of leaves. It is clearly unfounded. The evidence, wherever tested, shows a progressively reduced water loss from leaves resulting from a reduced water supply as the disease progresses. The evidence also indicates that the reduced water supply results from repeated extensions of vascular colonization by fungi followed by repeated extensions of host responses that seal off vessels, albeit too late to prevent colonization. In other words, successive barn doors are closed after the horse has escaped. By the time systemic infection has occurred, however, accumulated host products (IAA, ethylene, and numerous secondary metabolites resulting from belated defense responses), as well as accumulated metabolites of the pathogen, can also contribute to the complex of foliar symptoms and stunting commonly associated with "wilt" diseases.

Obviously, however, products of vascular fungi must play a role in disease development. Takai et al. (1983), using serological procedures, demonstrated that the fungal "toxin" cerato-ulmin, a small extracellular protein, was produced in situ in infected elms. This "toxin" has been shown to induce both internal, histological, as well as external, foliar symptoms comparable to those of the Dutch elm disease and to cause a reduction in the rate of transpiration from treated elm shoots. The question arises, does this molecule function directly to cause a failure of the water economy, or does it do so indirectly by eliciting host defense responses? In the final analysis, "toxins" and "elicitors" in the case of wilt disease, may turn out to be the same molecules — oligosaccharides, proteins or glycoproteins — that are involved with cell-to-cell recognition.

IX. PUTTING IT ALL TOGETHER

It is clear that contact with many organisms and events can trigger a cascade of defense responses in host cells. That is, recognition is generally non-specific, and that fact seems to be the basis for the relative suppressiveness of different soils. The number and variety of microorganisms in suppressive soils is high. The likelihood that a pathogenic form of a parasite will dominate at any infection site is therefore low. Since many, if not most, organisms trigger defense, resistance is the rule in suppressive soils. Furthermore, once triggered, these defense responses appear to have systemic effects that account for induced resistance.

The triggering of defense processes and resistance, in response to a chance vascular invader, is a common, every-day occurrence. Poor triggering of defense resulting in vascular colonization and disease is the exception. The conclusion follows that what is exceptional about a vascular pathogen is its ability (1) to go undetected or poorly detected (by mimicry?) or (2) actively to inhibit its recognition or response by the host (by means of a toxin?). It is also apparent, however, that resistance afforded by single-dominant "I-genes" is not absolute. The Q_{col} values (Table II) demonstrate that genetic control of resistance is expressed quantitatively and that there is a numerical continuum ranging from non-host resistance, through "I-gene-type" resistance, through various levels of polygenic resistance to the highly susceptible genetic condition demonstrated in the cultivar Bonny Best. These molecular signaling interactions, together with the transduction that follows, and their genetic control appear to be the factors that can bridge the gap between the genetic basis for resistance or susceptibility, on the one hand, and the degree of colonization (as measured by Q_{col}), on the other. These molecular interactions result in rapid or slow or essentially no progressive colonization and disease expression depending on the various genomes of plants, the soil types in which they grow and various soil management regimes employed. These are the basic biological facts that must be satisfied by any hypothesis for molecular interactions.

The use of single-dominant genes for resistance has been and continues to be the primary and preferred means for control of wilt diseases for many crop plants. Polygenic resistance has been relied upon where single-dominant genes are not available. The former is vulnerable to the evolution of new races of the pathogen and the latter to environmental conditions. In the future, plants will undoubtedly be "engineered" by the introduction of major genes for resistance, and this will be of unquestionable benefit. There is no guarantee, however, that the highly malleable genome of vascular pathogens will not lead to new pathogenic races that can by-pass or overcome such resistance. This danger can be mitigated by reducing the selection pressures that give rise to new races and allow them to build up. The use of soil management practices to provide soils of suppressive effect will clearly be beneficial and can serve to protect valuable resistance genes. The tailoring of soil nutrients to fit the situations of soil types and crop and potential pathogens will be as beneficial as tailoringthe genome. The use of both in combination can provide long-term stability in crop production.

REFERENCES

Alabouvette, C. (1989). Manipulation of soil environment to create suppressiveness in soils. *In* "Vascular Wilt Diseases of Plants" (E. C. Tjamos and C. H. Beckman, eds), pp. 457–478. NATO ASI Series H, Cell Biology, vol. 28. Springer-Verlag, Heidelberg.

Anonymous (1973). A Guide to the Use of Terms in Plant Pathology. Phytopathological Papers No. 17. Commonwealth Mycological Institute, Kew, Surrey, UK.

Atkinson, M. M. (1993). Molecular mechanisms of pathogen recognition by plants. *Advances in Plant Pathology* **10**, 35–64.

Awadalla, O. A. and El-Refai, I. M. (1992). Herbicide-induced resistance of cotton to Verticillium wilt disease and activation of host cells to produce the phytoalexin gossypol. *Canadian Journal of Botany* **70**, 1440–1444.

Beckman, C. H. (1987). "The Nature of Wilt Diseases of Plants". APS Press, St Paul, MN.

Beckman, C. H., Verdier, P. A. and Mueller, W. C. (1989). A system of defense in depth provided by vascular parenchyma cells of tomato in response to vascular infection with *Fusarium oxysporum* f. sp. *lycopersici*, race 1. *Physiological and Molecular Plant Pathology* **34**, 227–239.

Beckman, C. H., Morgham, A. T. and Mueller, W. C. (1991). Enlargement and vacuolization of the cytoplasm in contact cells in resistant and susceptible tomato plants following infection with *Fusarium oxysporum* f. sp. *lycopersici*, race 1. *Physiological and Molecular Plant Pathology* **38**, 433–442.

Benhamou, N., Joosten, M. H. A. J. and Wit, P. J. G. M. de (1990). Subcellular localization of chitinase and of its potential substrate in tomato root tissues infected by *Fusarium oxysporum* f. sp. *radicis-lycopersici*. *Plant Physiology* **92**, 1108–1120.

Bernards, M. A. and Ellis, B. E. (1991). Phenylalanine ammonia-lyase from tomato cell cultures inoculated with *Verticillium albo-atrum*. *Plant Physiology* **97**, 1494–1500.

Biles, C. L. and Martyn, R. D. (1989). Local and systemic resistance induced in watermelons by formae speciales of *Fusarium oxysporum*. *Phytopathology* **79**, 856–860.

Briggs, S. P. and Johal, G. S. (1994). Genetic patterns of plant host parasite interactions. *Trends in Genetics* **10**, 12–16.

Charchar, M. and Kraft, J. M. (1989). Response of near-isogenic pea cultivars to infection by *Fusarium oxysporum* f. sp. *pisi* races 1 and 5. *Canadian Journal of Plant Science* **69**, 1335–1346.

Cohen, R., Riov, J., Lisker, N. and Katan, J. (1986). Involvement of ethylene in herbicide-induced resistance to *Fusarium oxysporum* f. sp. *melonis*. *Phytopathology* **76**, 1281–1285.

Cohen, R., Yarden, O., Katan, J., Riov, J. and Lisker, N. (1987). Paclobutrazol and other growth-retarding chemicals increase resistance of melon seedlings to Fusarium wilt. *Plant Pathology* **36**, 558–564.

DeVay, J. E. (1989). Physiological and biochemical mechanisms in host resistance and susceptibility to wilt pathogens. *In* "Vascular Wilt Diseases of Plants" (E. C. Tjamos and C. H. Beckman, eds) pp. 196–217. NATO ASI Series H, Cell Biology, vol. 28. Springer-Verlag, Heidelberg.

DeVay, J. E. (1991). Historical review and principles of soil solarization. In "Soil Solarization" (J. E. DeVay, J. J. Stapleton and C. L. Elmore, eds) pp. 1–15. FAO Plant Production and Protection Paper No. 109, United Nations, Rome.

Dixon, R. A. and Lamb, C. J. (1990). Molecular communication in interactions between plants and microbial pathogens. *Annual Review of Plant Physiology and Plant Molecular Biology* **41**, 339–367.

Drøbak, B. K. (1993). Plant phosphoinositides and intracellular signaling. *Plant Physiology* **102**, 705–709.

Durrands, P. K. and Cooper, R. M. (1988). The role of pectinase in vascular wilt disease as determined by defined mutants of *Verticillium albo-atrum*. *Physiological and Molecular Plant Pathology* **32**, 363–371.

Elgersma, D. M. and Liem, J. I. (1989). Accumulation of phytoalexins in susceptible and resistant near-isolines of tomato inoculated with *Verticillium albo-atrum* or *Fusarium oxysporum* f. sp. *lycopersici*. *In* "Vascular Wilt Diseases of Plants" (E. C. Tjamos and C. H. Beckman, eds), pp. 237–246. NATO ASI Series H, Cell Biology, vol. 28, Springer-Verlag, Heidelberg.

Engelhard, A. W, Jones, J. P. and Woltz, S. S. (1989). Nutritional factors affecting Fusarium wilt incidence and severity. In "Vascular Wilt Diseases of Plants" (E. C. Tjamos and C. H. Beckman, eds), pp. 337–352. NATO ASI Series H, Cell Biology, vol. 28. Springer-Verlag, Heidelberg.

Ferraris, L., Gentile, I. A. and Matta, A. (1987). Activation of glycosidases as a consequence of infection stress in *Fusarium* wilt of tomato. *Journal of Phytopathology* **118**, 317–325.

Flor, H. H. (1942). Inheritance of pathogenicity in *Melampsora lini*. *Phytopathology* **32**, 653–659.

Gabriel, D. W., Loschke, D. C. and Rolfe, B. G. (1988). Gene-for-gene recognition: the ion channel defense model. *In* "Molecular Genetics of Plant–Microbe Interaction" (R. Palacios and D. Pal Verma, eds), pp. 3–14. APS Press, St Paul, MN.

Gao, H., Beckman, C. H. and Mueller, W. C. (1995a). Vascular colonization with wilt expression in susceptible or polygenically resistant or single-dominant-gene resistant tomato plants following inoculation with races 1, 2 or 3 of *Fusarium oxysporum* f. sp. *lycopersici* or with *F. o. pisi* or *cubense*. *Physiological and Molecular Plant Pathology* **46**, 29–43.

Gao, H., Beckman, C. H. and Mueller, W. C. (1995b). The nature of tolerance to *Fusarium oxysporum* f. sp. *lycopersici* in polygenically field-resistant Marglobe tomato plants. *Physiological and Molecular Pathology* **46**, 401–412.

Garibaldi, A., Brunatti, F. and Gullino, M. L. (1986). Suppression of Fusarium wilt of carnation by competitive non-pathogenic strains of *Fusarium*. *Mededelingen van de Faculteit Landbouwwelenschappen Rijksuniversiteit Gent* **51**, 633–638.

Gilroy, S. and Trewavas, A. (1990). Signal sensing and signal transduction across the plasma membrane. *In* "The Plant Plasma Membrane" (C. Larsson and I. M. Møller, eds), pp. 203–232, Springer-Verlag, Berlin and Heidelberg.

Gordon, T. R., Okamoto, D. and Jacobson, D. J. (1989). Colonization of muskmelon and nonsusceptible crops by *Fusarium oxysporum* f. sp. *melonis* and other species of *Fusarium*. *Phytopathology* **79**, 1095–1100.

Hadwiger, L. A. (1991). Nonhost resistance in plant fungal interactions. *In* "Molecular Signals in Plant Microbe Communications" (D. Pal Verma ed.), pp. 85–96. CRC Press, Boca Raton, FL.

Hahn, M. G., Cheong, J.-J., Alba, R. and Côté, F. (1993). Oligosaccharide elicitors: structures and signal transduction. *In* "Plant Signals in Interactions with Other Organisms" (Current Topics in Plant Physiology, vol. 11) (J. Schultz and I. Raskin, eds), pp. 24–46. American Society of Plant Physiologists, Rockville, MD.

Harrison, N. A. and Beckman, C. H. (1987). Growth inhibitors associated with Fusarium wilt of tomato. *Physiological and Molecular Plant Pathology* **30**, 401–420.

Huisman, O. C. and Gerik, J. S. (1989). Dynamics of colonization of plant roots by *Verticillium dahliae* and other fungi. *In* "Vascular Wilt Diseases of Plants" (E. C. Tjamos and C. H. Beckman, eds), pp. 1–17. NATO ASI Series H, Cell Biology, vol. 28. Springer-Verlag, Heidelberg.

Jordan, C. M., Endo, R. M. and Jordan, L. S. (1988). Association of phenol-containing structures with *Apium graveolens* resistance to *Fusarium oxysporum* f. sp. *apii*, race 2. *Canadian Journal of Botany* **66**, 2385–2391.

Jorge, P. E., Green, R. J., Jr and Chaney, W. R. (1992). Inoculation with *Fusarium* and *Verticillium* to increase resistance in *Fusarium*-resistant tomato. *Plant Disease* **76**, 340–343.

Katan, J. (1989). Soil temperature interactions with the biotic components of vascular wilt diseases. *In* "Vascular Wilt Diseases of Plants" (E. C. Tjamos and C. H. Beckman, eds), pp. 353–366. NATO ASI Series H, Cell Biology, vol. 28. Springer-Verlag, Heidelberg.

Kauss, H. (1990). Role of the plasma membrane in host–pathogen interactions. *In* "The Plant Plasma Membrane" (C. Larsson and I. M. Møller eds), pp. 320–350. Springer-Verlag, Berlin and Heidelberg.

Keen, N. T. (1990). Gene-for-gene complementarity in plant pathogen interactions. *Annual Review of Genetics* **24**, 447–463.

Krebs, S. L. and Grumet, R. (1993). Characterization of celery hydrolytic enzymes induced in response to infection by *Fusarium oxysporum*. *Physiological and Molecular Plant Pathology* **43**, 193–208.

Lamb, C. J., Lawton, M. A., Dron, M. and Dixon, R. A. (1989). Signals and transduction mechanisms for activation of plant defenses against microbial attack. *Cell* **56**, 215–224.

Lemanceau, P. (1989). Role for competition for carbon and iron in mechanisms of soil suppressiveness to Fusarium wilts. *In* "Vascular Wilt Diseases of Plants" (E. C. Tjamos and C. H. Beckman, eds), pp. 388–396. NATO ASI Series H, Cell Biology, vol. 8. Springer-Verlag, Heidelberg.

Lemanceau, P., Alabouvette C. M. and Couteaudier, Y. (1988). Recherches sur la résistance des sols aux maladies. XIV. Modifications du niveau de réceptivité d'un sol résistant et d'un sol sensible aux fusarioses vasculaires en réponse à des apports de fer ou de glucose. *Agronomie* **8**, 155–162.

Louvet, J. (1989). Microbial populations and mechanisms determining soil-suppressiveness to Fusarium wilts. *In* "Vascular Wilt Diseases of Plants" (E. C. Tjamos and C. H. Beckman, eds), pp. 367–384. NATO ASI Series H, Cell Biology, vol. 28. Springer-Verlag, Heidelberg.

Mace, M. E. (1989). Secondary metabolites produced in resistant and susceptible host plants in response to fungal vascular infection. In "Vascular Wilt Diseases of Plants" (E. C. Tjamos and C. H. Beckman, eds), pp. 163–174. NATO ASI Series H, Cell Biology, vol. 28. Springer-Verlag, Heidelberg.

Mace, M. E., Stipanovic, R. D. and Bell, A. A. (1985). Toxicity and role of terpenoid phytoalexins in Verticillium wilt resistance in cotton. *Physiological and Molecular Plant Pathology* **26**, 209–218.

Martin, G. B., Brommonschenkel, S. H., Chunwongse, J., Frary, A., Ganal, M. W., Spivey, R., Wu, T., Earle, E. D. and Tanksley, S. D. (1993). Map-based cloning of a protein kinase gene conferring disease resistance in tomato. *Science* **262**, 1432–1436.

Matta, A. (1989). Induced resistance to Fusarium wilt diseases. In "Vascular Wilt Diseases of Plants" (E. C. Tjamos and C. H. Beckman, eds), pp. 175–196. NATO ASI Series H, Cell Biology, vol. 28. Springer-Verlag, Heidelberg.

Mauseth, J. D. (1988). "Plant Anatomy". Benjamin/Cummings, Menlo Park, CA.

Moreau, M., Catesson, A. M., Peresse, M. and Czaninski, Y. (1978). Dynamique comparée des réactions cytologiques du xylème de l'oeillet en présence de parasites vasculaires. *Phytopathologische Zeitschrift* **91**, 289–306.

Mueller, W. C., Morgham, A. T. and Roberts, E. M. (1994). Immunocytochemical localization of callose in the vascular tissue of tomato and cotton plants infected with *Fusarium oxysporum*. *Canadian Journal of Botany* **72**, 505–509.

Newcombe, G. and Robb, J. (1989). The chronological development of a lipid-to-suberin response at *Verticillium* trapping sites in alfalfa. *Physiogical and Molecular Plant Pathology* **34**, 55–73.

Newcombe, G., Papadopoulos, Y. A., Robb, J. and Christie, B. R. (1989). The colonization ratio: a measure of the pathogen invasiveness and host resistance in Verticillium wilt of alfalfa. *Canadian Journal of Botany* **67**, 365–370.

Nicholson, R. L. and Hammerschmidt, R. (1992). Phenolic compounds and their role in disease resistance. *Annual Review of Phytopathology* **30**, 369–389.

Niemann G. J., Bij, A. van der, Brandt-de Boer, B., Boon, J. J. and Baayen, R. P. (1991a). Differential responses of four carnation cultivars to races 1 and 2 of *Fusarium oxysporum* f. sp. *dianthi* and to *Phialophora cinerescens*. *Physiological and Molecular Plant Pathology* **38**, 117–136.

Niemann, G. J., Kerk, van der, A., Niessen, W. M. A. and Versluiis, K. (1991b). Free and cell wall-bound phenolics and other constituents from healthy and fungus-infected carnation (*Dianthus caryophyllus* L.) stems. *Physiological and Molecular Plant Pathology* **38**, 417–432.

Peer, R. van, Neimann, G. J., and Schippers, B. (1991). Induced resistance and phytoalexin accumulation in biological control of Fusarium wilt of carnation by *Pseudomonas* sp. strain WCS417r. *Phytopathology* **81**, 728–734.

Pennypacker, B. W. and Leath, K. T. (1993). Anatomical response of resistant alfalfa infected with *Verticillium albo-atrum*. *Phytopathology* **83**, 80–85.

Price, D. and Sackston, W. E. (1989). Cross protection among strains of *Verticillium dahliae* in sunflower. In "Vascular Wilt Diseases of Plants" (E. C. Tjamos and C. H. Beckman, eds), pp. 229–235. NATO ASI Series H, Cell Biology, vol. 28. Springer-Verlag, Heidelberg.

Robb, J., Lee, S. W., Mohan, R. and Kolattukudy, P. E. (1991). Chemical characterization of stress-induced vascular coating in tomato. *Plant Physiology* **97**, 528–536.

Rouxel, F., Alabouvette, C. and Masson, J. P. (1988). Intéret de la modélisation pour l'etude de la réceptivité des sols aux maladies d'origine tellrique: examples d'application a la hemie des cruciféres et aux fusarioses vasculaire. *Comptes Rendus Academe Agricole Française* **73**, 137–149.

Ryan, C. A. and Farmer, E. E. (1991). Oligosaccharide signals in plants: a current assessment. *Annual Review of Plant Physiology and Molecular Biology* **42**, 651–674.

Sanders, L., Wang, C.-O., Walling, L. and Lord, E. (1991). A homolog of the substrate adhesion molecule vitronectin occurs in four species of flowering plants. *Plant Cell* **3**, 629–635.

Schneider, R. W. (ed.) (1982). "Suppressive Soils and Plant Disease". American Phytopathological Society, St Paul, MN.

Schneider, R. W. (1985). Suppression of Fusarium yellows of celery with potassium, chloride and nitrate. *Phytopathology* **75**, 40–48.

Segal, G., Sarfatti, M., Schaffer, M. A., Ori, N., Zamir, D. and Fluhr, R. (1992). Correlation of genetic and physical structure in the region surrounding the I_2 *Fusarium oxysporum* resistance locus in tomato. *Molecular and General Genetics* **231**, 179–185.

Shi, J., Mueller, W. C. and Beckman, C. H. (1991a). Ultrastructure and histochemistry of lipoidal droplets in vessel contact cells and adjacent parenchyma cells in cotton plants infected by *Fusarium oxysporum* f. sp. *vasinfectum*. *Physiological and Molecular Plant Pathology* **38**, 201–211.

Shi, J., Mueller, W. C. and Beckman, C. H. (1991b). Ultrastructural responses of vessel contact cells in cotton plants resistant or susceptible to infection by *Fusarium oxysporum* f. sp. *vasinfectum*. *Physiological and Molecular Plant Pathology* **38**, 211–222.

Shi, J., Mueller, W. C. and Beckman, C. H. (1992). Vessel occlusion and secretory activities of vessel contact cells in resistant and susceptible cotton plants infected with *Fusarium oxysporum* f. sp. *vasinfectum*. *Physiological and Molecular Plant Pathology* **40**, 133–147.

Shi, J., Mueller, W. C. and Beckman, C. H. (1993). The inhibition of fungal growth in resistant cotton plants infected by *Fusarium oxysporum* f. sp. *vasinfectum*. *Journal of Phytopathology* **139**, 253–260.

Stein, J. C., Howlett, B., Boyes, D. C., Nasrallah, M. E. and Nasrallah, J. B. (1991). Molecular cloning of a putative receptor protein kinase gene encoded at the self-incompatibility locus of *Brassica oleracea*. *Proceedings of the National Academy of Sciences USA* **88**, 8816–8820.

Sutherland, M. L. and Pegg, G. F. (1992). The basis of host recognition in *Fusarium oxysporum* f. sp. *lycopersici*. *Physiological and Molecular Plant Pathology* **40**, 423–436.

Takai, S., Richards, W. C. and Stevenson, K. J. (1983). Evidence for the involvement of cerato-ulmin, the *Ceratocystis ulmi* toxin, in the development of Dutch elm disease. *Physiological Plant Pathology* **23**, 275–280.

Trewavas, A. and Gilroy, S. (1991). Signal transduction in plant cells. *Trends in Genetics* **7**, 356–361.

Vander Molen, G. E., Labavitch, J. M., Strand, L. L. and DeVay, J. E. (1983). Pathogen-induced vascular gels: ethylene as a host intermediate. *Physiologia Plantarum* **59**, 573–580.

Vander Molen, G. E., Labavitch, J. M. and DeVay, J. E. (1986). Fusarium-induced vascular gels from banana root – a partial chemical characterization. *Physiologia Plantarum* **66**, 298–302.

Vander Molen, G. E., Beckman, C. H. and Rodehorst, E. (1987). The ultrastructure of tylose formation in resistant banana following inoculation with *Fusarium oxysporum* f. sp. *cubense*. *Physiological and Molecular Plant Pathology* **31**, 185–200.

Wagner, V. T. and Matthysse, A. G. (1992). Involvement of a vitronectin-like protein in attachment of *Agrobacterium tumefaciens* to carrot suspension culture cells. *Journal of Bacteriology* **174**, 5999-6003.

Walker, J. C. (1971). "Fusarium Wilt of Tomato". Monograph 6. American Phytopathological Society, St Paul, MN.

Weiler, E. W. (1993). Octadecanoid-derived signaling molecules involved in touch perception in a higher plant. *Botanica Acta* **106**, 2-4.

Zhang, J., Mace, M. E., Stipanovic, R. D. M. and Bell, A. A. (1993). Production and fungitoxicity of the terpenoid phytoalexins in cotton inoculated with *Fusarium oxysporum* f. sp. *vasinfectum*. *Journal of Phytopathology* **139**, 247-252.

Implications of Population Pressure on Agriculture and Ecosystems

ANNE H. EHRLICH

Department of Biological Sciences, Stanford University, Stanford, California 94305, USA

I. INTRODUCTION

Today, as nations emerge from the shadows of the cold war, and as economic connections and modern communications knit diverse societies into a truly global civilization, we find ourselves face to face with a fundamental question: just how many people can be supported on Earth? On entering the 21st

Advances in Botanical Research Vol. 21
incorporating Advances in Plant Pathology
ISBN 0-12-005921-5

century, the human population will number over 6 billion and will still be grow-ing. The United Nations' most recent medium projection indicates continued growth for another century or so to a population size of 11.6 billion — more than twice the 1994 size of 5.6 billion (United Nations Population Division, 1991).

Can so many human beings be supported on Earth, a planet of finite size with finite resources? For how long? What are the prospects for greatly increasing food production? What limiting factors might be encountered? What environmental costs will be incurred in the process of such an expan-sion? What choices and tradeoffs will civilization face in the attempt to support many billions more people? How might a transition to a more secure world than today's be managed without disaster?

II.   CARRYING CAPACITY, TECHNOLOGY, AND SUSTAINABILITY

These questions cannot be addressed without first considering Earth's carrying capacity for human life and the relationship of the human population to it. Biologists define carrying capacity as the maximum number of individuals of an organism that can be supported indefinitely without degrading the capacity of their environment to support future populations. Carrying capacity is thus a function both of the characteristics of the environment — the resources it contains and its productivity, and of the organism itself — its physical size and the amounts of energy and resources each individual requires to meet its needs (Daily and Ehrlich, 1992). A population living within its carrying capacity thus is sustainable over the long term (to use the buzzword of the moment). A population that substantially exceeds its carrying capacity for any length of time will deplete its resource base and/or degrade the environment's produc-tive capacity. The usual consequences are a sharp rise in the death rate and often the departure of many individuals for greener pastures.

Carrying capacity is easily defined, but calculating it for any organism is a complex matter. The concept also applies to human beings, of course, although they are by no means typical organisms. People, like other animals, need oxygen, water, food to provide energy and essential nutrients, shelter, living space, a hospitable environment, and contact with others of their kind. But human beings, far more than other animals, also use materials to enhance their environments and gain access to other resources. Human technologies can dramatically change an area's carrying capacity, particularly in the short to medium term. Technology enables people to manipulate biotic resources to increase the production of desired organisms (such as crops and livestock), to obtain resources (such as metals and fossil fuels) that are inaccessible to other animals, and to trade materials over large distances.

The result of this ability has been in effect to increase manyfold the carrying

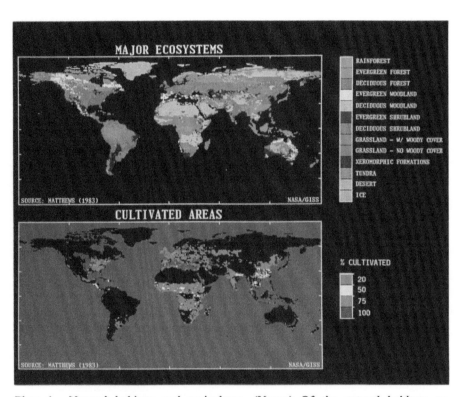

Plate 1. Natural habitats and agriculture. (Upper) Of the natural habitats on Earth's land surface, the most productive (blessed with fertile soils and favorable climate, including adequate moisture) generally are grasslands, woodlands, shrublands, and some forest areas. (Lower) The red colored areas show the extent to which agriculture has replaced natural habitat on Earth. Compliments of Elaine Matthews, NASA Goddard Institute for Space Studies, New York. For data see Matthews (1983).

capacities for human beings of many regions of the world. Over ten millennia, a population of only a few million hunter/gatherers expanded into an agricultural population more than two orders of magnitude larger; it increased six-fold more in less than two centuries since the dawn of the industrial age. This enormous expansion in the human enterprise was supported by a commensurate increase in food production and an accelerating mobilization of physical resources. In the process, humanity has made dramatic changes in Earth's land surface (Turner *et al.*, 1990), diverting an ever larger share of the productivity of the biosphere to human uses (Vitousek *et al.*, 1986), and has begun to alter natural biogeochemical processes. With the increasing takeover of the world's land, civilization has launched a nearly unprecedented epidemic of biotic extinctions (Wilson, 1989; Ehrlich and Wilson, 1991), with largely unpredictable consequences for humanity's future carrying capacity (Ehrlich and Ehrlich, 1981).

Yet, although ingenuity has enormously increased the human carrying capacity, especially in the last century, growth of the human enterprise in the end will none the less be constrained by inescapable biophysical limits. The daily income to Earth of solar energy is limited, as is the amount of accessible energy from past solar income stored in the form of fossil fuels. There are also limits to environmental tolerances for abuse and manipulation, although these are less easily assessed. Most of the limiting factors, of course, are likely to manifest themselves first and foremost as constraints to increasing food production.

And, despite humanity's technological accomplishments, social constraints and inefficiencies (such as lifestyles, land use practices, and economic arrangements) in effect reduce carrying capacities. These presumably will always be with us (Keyfitz, 1991), just as physical limits such as the laws of thermodynamics will be. Perhaps we should be grateful. Increasing the "social carrying capacity" to match the potential biophysical carrying capacity might lead to the maximum number of people being supported at a minimal standard of living: the human equivalent of the factory farming of chickens (Daily and Ehrlich, 1992).

The carrying capacity for human beings is thus surprisingly difficult to calculate, and the result, to a large extent, depends on the choice of lifestyles and levels of resource use. Regardless of chosen living standards, however, an essential aspect of the carrying capacity concept is that the population must be sustainable over the long term.

The term "sustainable" has caused considerable confusion in recent years while gaining widespread use. In 1987, the World Commission on Environment and Development, headed by Gro Haarlem Brundtland, of Norway, defined "sustainable development" as follows: "to ensure that [development] meets the needs of the present without compromising the ability of future generations to meet their own needs". (World Commission on Environment and Development, 1987.).

The sudden interest in sustainability is timely, inasmuch as symptoms of population "overshoot" (exceeding the carrying capacity) are already increasingly evident to biologists, who view most societies as living beyond their means. Humanity is doing what no responsible family or corporation would do: consuming capital to support itself (Ehrlich and Ehrlich, 1991).

The capital in question is the world's endowment of resources, both renewable and non-renewable. That civilization is depleting reserves of non-renewable resources such as metals and fossil fuels (especially petroleum) is well known. This causes little concern, largely because many economists believe that acceptable substitutes can always be found for any resource (Daly and Cobb, 1989). Much less appreciated, but more worrisome, is the depletion of putatively renewable resources such as agricultural soils (Brown and Wolf, 1984; Brown *et al.*, 1994a), fossil groundwater (Postel, 1990), and biodiversity (Ehrlich and Ehrlich, 1981; Wilson, 1989), all of which are particularly essential for food production. Since there are no known substitutes for any of these resources, their disappearance is obviously unsustainable.

III.   POPULATION GROWTH

A continuously growing population in a finite environment is by definition unsustainable over the long term. In 1994, the world population passed 5.6 billion and continues to increase by almost 90 million people per year (Haub, 1992, 1994; Haub and Yanagishita, 1994). The rate of growth (births minus deaths) was about 1.6% per year, which would double the population in 43 years if it continued unchanged. Indeed, the world population has nearly tripled since 1930 − in less than an average lifetime.

The United Nations medium demographic projection of 1991 indicated that the population is likely to pass 8.5 billion around 2025, reach 10 billion near mid-century, and ultimately stop at 11.6 billion (United Nations Population Division, 1991; United Nations Population Fund, 1992). Fig. 1 shows the United Nations' medium projection broken down by region, with all developed nations combined in the segment at the bottom. This is the UN's "most likely" prognostication, but unexpected changes in reproductive or mortality rates, of course, could cause growth to be either considerably faster or slower.

The UN's high projection shows no end to growth, with the population soaring past 28 billion around 2150 − a clearly unsustainable path. The low projection indicates a peak population size of 8.5 billion around 2050, followed by a slow decline, passing 5 billion again within a century. The key assumption here is that the average completed family size worldwide (now 3.2 children per couple) would soon fall below "replacement reproduction" − the level at which parents on average just replaced themselves in the next generation: about 2.1 children per couple (McNicoll, 1992).

The low path presupposes major changes both in national policies and

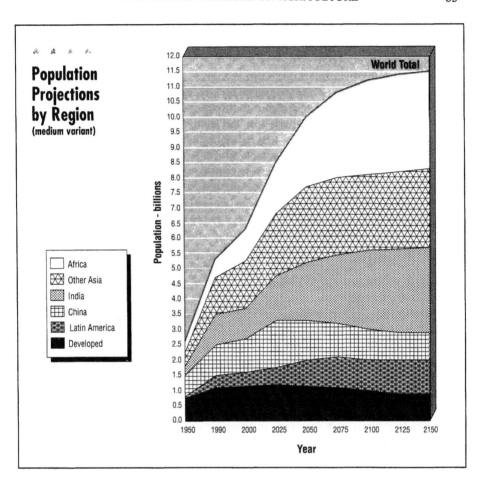

Fig. 1. Population projections by region (medium variant). The medium variant is considered the most likely. Source: United Nations Population Fund.

individual reproductive behavior; that is, a marked strengthening of population programs and policies in most nations and changed motivations among couples toward smaller families. Today, although most industrialized nations lack formal population policies, they have low birthrates (most are below replacement level), low to zero or even negative population growth rates, and adequate access to means of birth control. Developing nations in the 1990s tend to have higher birth and growth rates, although their total fertility rates (TFR: the average number of children each woman in a society will bear in her lifetime at current age-specific fertility rates) vary from fewer than two to more than seven children per woman. With the latter TFR (and with low mortality rates), a population could double in as little as 18 years. Access to family planning services also varies among developing countries from excellent to negligible, even though most countries have government-supported family planning programs (UNFPA, 1992).

Achievement of replacement reproduction or less is a process that usually takes decades when TFRs are initially high. Even if a worldwide TFR of 2.1 could be achieved soon (say, within two decades), the momentum of population growth — caused by the preponderance of young people in a previously rapidly growing population — will delay the end of growth. If a population exactly achieves replacement reproduction but no less, it will take about 70 years, roughly a lifetime, to end growth. Lower fertility will end growth sooner and lead to negative growth (population shrinkage), but there must still be a lag of several decades. Thus the United States, which has had below-replacement reproduction for two decades, still has an annual rate of natural increase (births minus deaths) of 0.7%. Germany, on the other hand, has wrung momentum out of its population structure and achieved a slight negative growth rate, as have several Eastern European nations, including Russia (Haub and Yanagishita, 1994).

Population momentum explains why, even if unexpectedly rapid reductions in fertility occurred in nations with high birth rates, the world population is destined to grow substantially further. One third of today's 5.6 billion — almost 2 billion — are under age 15, whereas fewer than 300 million are over 65. The youngsters are the parents of the next generation, and will at least double their own numbers, whereas the old folks who die in the next few decades will make scarcely a dent. Given present population structures, the projected population increase of 3–8 billion within the next century could only be avoided by a massive rise in death rates. This possibility is seldom acknowledged by demographers and never projected because there is no basis for prediction (with the notable exception of projected increases in death rates from AIDS in heavily affected regions such as Central Africa).

IV. FEEDING THE BILLIONS

Whether the world's demographic future will actually resemble any of the United Nations' projections depends on humanity's ability to support for many decades a population much larger than today's. Maintaining the food supply is probably the most critical component of that support. Several times in the last half-century, the question has been raised as to whether and for how long the rapidly expanding human population can maintain an adequate food supply. The question has not much been heard for the last decade, but clearly it deserves serious attention (Paddock, 1990, 1992).

With continued population growth, raising food production to keep pace becomes increasingly problematic (Brown et al., 1994) as pressures rise on the resources and natural processes underpinning agricultural activities (Stavens,

1994). These include cultivable land with fertile soils, dependable supplies of fresh water, and the healthy functioning of natural ecosystems in which agricultural systems are embedded (Pimentel and Hall, 1989).

Nearly all the potentially farmable land in the world — land that is reasonably level and has a suitable climate and productive soil — is already being cultivated. From 1950 to 1990, worldwide production of grains (the principal feeding base of humanity, supplying roughly half the calories consumed) nearly tripled, while the population a little more than doubled. This impressive gain in food production, however, was achieved mainly by intensifying cultivation on existing farmland, not by substantially increasing the area of land under crops; the world's cropland area (including that planted in grains, pulses, and tubers) increased by only about 25% (Grigg, 1993). In recent decades, increasing amounts of farmland have been taken out of production for several reasons: expanding urban development, which is often built over prime farmland; depletion of topsoil; or degradation of land due to poor irrigation practices (Brown et al., 1994a).

Modern farming requires a generous fossil fuel subsidy to support the high crop yields and intensive practices necessary to feed the present population. Fossil fuels are used both to provide raw materials and fuel the production of pesticides and fertilizers; they are also needed to power farm machinery and irrigation works, as well as to transport crops to market and inputs (seed, farm chemicals) to the farm. Given that accessible reserves of fossil fuels, especially petroleum, are limited and expected to run short within a few decades (oil) to centuries (coal), this dependence is patently unsustainable (Holdren, 1990).

The use of the seeds of specially bred, high-yielding varieties of major cereal crops, together with high rates of fertilizer applications, pesticides, and abundant water (in order to achieve the high yields), is known as the green revolution (Dahlberg, 1979; Swaminathan and Sinha, 1986). The green revolution began in Western nations and was spread to much of the developing world during the 1960s and 1970s. The doubling and tripling of yields of a few staple crops — mainly wheat, rice and maize — allowed world food production to increase faster than the population from 1950 until the mid-1980s. Production increases were also achieved in other crops, such as root crops (which are important dietary components in some regions, especially sub-Saharan Africa), but not to the extent that was realized in the major cereals (Grigg, 1993).

Since 1985, however, global production of grains has lagged on a per-capita basis (Brown et al., 1994a; USDA, 1994). The green revolution technology has now been applied in most suitable regions and will soon have run its course. In most areas, diminishing returns from increased fertilizer applications are already apparent (Brown, 1991). Although continued breeding of new high-yield varieties of major grain crops still leads to yield improvements in most of them, rice breeders have in recent years encountered what seems to be a "yield cap" (Walsh, 1991; IRRI, 1992). In experimental plots, the yields of the newest rice strains, although exhibiting desirable traits such as resistance to important pests, do not match those of some earlier varieties. Sooner or later the same problem is likely to appear in other crops (Bugbee and Monje, 1992).

Furthermore, no technology can compensate fully for the loss of arable land or for depleted or damaged soils, although generous fertilizer applications can mask the effects of depletion for some time (Brown, 1988). Ironically, the green revolution itself, although dramatically boosting food harvests, often accelerates the deterioration of agricultural lands. Continuous cultivation, abetted by synthetic fertilizers, speeds soil erosion and nutrient loss. Green-revolution crops are normally planted in large-scale monocultures (to facilitate use of farm machines), which invite pest or crop disease attacks, unlike the shifting mosaics of small-scale mixed crop cultures characteristic of traditional agriculture, especially in the tropics (Wrigley, 1982; Altieri, 1983). Heavy applications of fertilizers and pesticides often cause serious pollution problems, especially in surface and groundwater sources. They also pose threats to human health, especially for farm workers who use the pesticides (World Resources Institute *et al.*, 1994). Farm chemicals are known to alter soil chemistry (World Resources Institute *et al.*, 1992), but little is known about their impacts on soil organisms, despite the importance of those organisms in maintaining soil fertility (Hillel, 1991).

Most of the problems of the green revolution are intensified when attempts are made to transfer the technology to moist tropical areas (Ehrlich, 1994). Where there is no winter or marked dry season to reduce insect pest populations or crop diseases, their attacks are relentless; continuous planting of the same crop in a large-scale monoculture is simply impractical in such situations. Torrential tropical rains quickly leach out nutrients (including those applied in fertilizers), and wash away pesticides. Efforts to compensate by using heavier and more frequent chemical applications increase the pollution problems in soil and surface waters. Moreover, tropical soils are often nutrient-poor and thin, which makes them ill suited to continuous cultivation (Sanchez, 1976). Thus attempts to introduce Western-style modern agriculture in the humid tropics have often failed (Goodland *et al.*, 1984; Fearnside, 1987).

Exceptions to these general observations appear to be in areas such as Java in Indonesia, where rich volcanic soils do permit continuous cultivation, although when attempts were made to transfer large numbers of colonists from Java to other, less well-endowed islands in the archipelago, the problems appeared. Tropical Asian rice-growing regions in general, however, have had much greater success with green revolution technologies than have attempts with other crops in Latin America and Africa. Rice cultivation differs from dry grain production in several respects. Huge monocultures are relatively uncommon, given small farm sizes and the separation of paddies. Draining and drying of paddies between crops interrupts the build-up of pest populations, and the custom of combining aquaculture with rice-growing also helps control pests. Weeds have traditionally been controlled by hand, although in several countries farmers have begun switching to the use of herbicides as urban demand for labor increases farm labor costs (Naylor, 1994).

Yet no encore for the green revolution is in sight. Although biotechnology is often touted as the next green revolution, its potential for expanding food supplies is comparatively limited. Judging from current research trends, the value of biotechnology will lie more in improving market quality and shelf life of some foods, providing better inbred resistance to diseases and pests, and perhaps increasing the efficiency of utilization by crops of particular nutrients or photosynthetic efficiency. Most of the work today, moreover, seems to be concentrated on cash crops consumed mainly by relatively affluent people, such as tomatoes, tropical fruits, coffee, and cotton (Gasser and Fraley, 1989; World Resources Institute *et al.*, 1994). Harvests may be marginally increased and security against losses caused by pests somewhat enhanced, but another doubling of yields in already high-yielding staple crops is unlikely, at least in the medium term.

V. AGRICULTURE'S RESOURCE BASE

The primary limiting resource for agriculture is obviously productive land, a resource often claimed to be in abundant supply by technological optimists who may overlook either physical limitations or social constraints on that supply (FAO *et al.*, 1982; Lee *et al.*,1988; Swaminathan and Sinha, 1986). Yet most of the land that is suitable for farming is already under cultivation. Plate 1 is of satellite images showing the different natural habitat types on Earth and the extent to which agriculture has taken over native habitat (see also Matthews, 1983). Much of human history has been one of expansion of land under the plow, but this expansion all but ended by about 1970. Even though new land has continued to be opened, mainly in tropical regions and often as a result of deforestation (Myers, 1989), other land has increasingly come out of production. In the humid tropics, much deforested land could be successfully cultivated for only a few years, then it was demoted to pasture for a few more years before abandonment. Desertification, severe erosion, lack of water, and irrigation damage have led to abandonment of crop land in many areas of the world. Sometimes marginal land has been removed from cultivation for soil conservation purposes, as in the United States starting in 1985 (Brown, 1988). Finally, an increasing amount of good farm land is being lost every year to road-building and urbanization.

At least as important as the quantity of land available for food production is its quality. However, in the last half-century, as the world's crop land area has essentially stopped increasing, its overall quality has also been declining at a rate that is far from encouraging. The amount of topsoil estimated to be lost from erosion each year from the world's agricultural land during the 1980s was roughly 24 billion tons, a rate that is far from sustainable even in the medium term (Brown and Wolf, 1984). Although, estimates of worldwide losses are extremely difficult to make, since erosion rates vary enormously from place to place, it is safe to say that erosion is commonplace and is occurring on average many times faster than rates of natural replenishment (Pimentel, 1993). Annual losses of 10–20 tonnes/ha of farm land are normal in North America and Europe (as opposed to about one tonne of replenishment), and in many tropical regions, erosion of hundreds of tonnes per hectare are not unusual.

A more detailed study by the United Nations Environment Programme has concluded that significant degradation of the world's productive land has occurred since 1945 on every continent but Antarctica (Oldeman *et al.*, 1990; World Resources Institute *et al.*, 1992). Cropland, forest land, and grasslands have all been significantly damaged by overintensive agriculture, deforestation, and overgrazing. Eleven percent of the world's productive lands were assessed as moderately to extremely damaged by 1990, and a further 6% lightly degraded and at risk of further deterioration.

The fraction of land degraded varies among the world's regions, with Central America and Mexico having suffered at least moderate deterioration on 24% of productive lands (e.g. Maass and Garcia-Oliva, 1992), whereas North America and Oceania had moderate or worse damage on less than 5%. The leading types of degradation in the UNEP study, accounting for more than three quarters worldwide, were water and wind erosion, followed by nutrient depletion, loss of native vegetative species (in rangelands), salination of soil, and loss of soil structure. Moderate to severe damage causes measurable losses of productivity, such as reduced crop yields or livestock-carrying capacities. Restoration of productivity to previous levels would be difficult and costly, and extremely degraded lands were deemed beyond reclamation.

Fresh water is already in seriously short supply in many regions, not only for food production but in some areas even for domestic use (Gleick, 1993; Falkenmark and Widstrand, 1992). Water has become a major limiting factor for development in many areas, often those with rapid population growth, such as the Middle East, northern India, and much of Africa, and also in China (Tyler, 1993).

Irrigation has been an essential ingredient in the near-tripling of global food production since 1950, and much of the water came from groundwater sources (Postel, 1990). Today an increasing amount of irrigated farmland is being taken out of production in many parts of the world, including Russia, the United States, the Middle East, India, and China. Irrigation is being abandoned either because accumulations of salt or waterlogging have reduced productivity or because depletion of aquifers has made withdrawals uneconomic. To support high-yield agriculture, aquifers in many areas have been drained at rates far above those of natural recharge; an important example is the Ogallala Aquifer underlying the US great plains. Many groundwater sources have also been contaminated with toxic substances and thus rendered unusable for agriculture or domestic consumption.

An important agricultural resource is the genetic diversity that exists in the dozens of thousands of different traditional varieties of each significant crop species, which have been bred for various characteristics by farmers for millennia (Hoyt 1988; National Research Council, 1991). The introduction of the green revolution technology, with a severely limited array of varieties of each crop, has led to farmers abandoning their traditional crop strains, with the result that perhaps hundreds have been lost forever (Dahlberg, 1979). But the genetic variability represented in traditional strains and in the wild relatives of crops (many of which are seriously endangered around the world) is the necessary raw material for breeding new, improved varieties of crops (Plunknett *et al.*, 1987). It is also essential for staying ahead in the coevolutionary race with pests and crop diseases, which continuously evolve ways to overcome both natural resistance and pesticides (Ehrlich and Ehrlich, 1991). Establishment of gene banks in many regions of the world has at least slowed the loss of traditional varieties, but there are many problems with retaining viability and vigor in stored seeds (Cohen *et al.*, 1991). Long-term preservation can only be assured by regular planting and harvesting.

VI. ECOSYSTEMS AND AGRICULTURE

The component of earthly capital being heedlessly destroyed that is probably least appreciated by the general public is biodiversity — the complement of plants, animals, and micro-organisms with which we share the planet (and are possibly our only living companions in the universe). Earth's land areas have been increasingly taken over and modified for direct human occupation, degraded by overintensive agriculture, overgrazing, and large-scale deforestation, and assaulted by human-released toxins from air pollutants and acid precipitation to pesticides. Natural ecosystems almost everywhere have been fragmented, disrupted, or destroyed. Consequently, habitats for other organisms have disappeared, leading to an accelerating epidemic of extinctions (Ehrlich and Ehrlich, 1981; Wilson and Peter, 1988; Ehrlich and Wilson, 1991). Biologists estimate that perhaps a quarter of the original number of species on Earth may disappear by 2025.

Natural ecosystems provide a series of essential services to humanity that are especially vital for agriculture (Ehrlich and Ehrlich, 1981). When ecosystems are degraded or destroyed outright, the provision of these amenities may be impaired or lost. Among the ecosystem services are: maintenance of the gaseous composition of the atmosphere; moderation of climate and regulation of the hydrological cycle; cycling of nutrients, replenishment of soils, and disposal of wastes; pollination (including of numerous crops); control of the vast majority of pests and disease vectors; detoxification of toxic substances; direct provision of forest products, seafood and freshwater fishes, and innumerable other foods and materials taken directly from nature.

Individual species could be the source or inspiration for countless potential new products, from foods, medicines, and spices to industrial chemicals, all to be found in nature's vast genetic library — if the library is not first dismantled. The decimation of species is serious enough, but the loss of genetically distinct populations, from the standpoint of ecosystem services and potentially useful products, is even more critical.

Most ecosystem services are taken for granted and valued only after being lost (Ehrlich and Mooney, 1983; Ehrlich and Ehrlich, 1992; Ehrlich, 1993). Even if people knew how to replace these services (in most cases they do not), replacement by technological means on the required scale would be impossible. Attempts to replace ecosystem services (such as management of the hydrological cycle through dams and water projects) are far from perfect and often unsustainable.

That civilization is imposing profound changes on the world's ecosystems can be seen in the aggregate human impact on biological productivity — the most fundamental resource of all, the basis of all food for human beings and other animals. Net primary production (NPP) is the energy from sunlight made available through the process of photosynthesis by green plants, algae, and some kinds of bacteria (excepting the energy those organisms use for their own life processes). People directly consume as food, livestock feed, fiber, and forest products about 5.5% of the total amount of energy made available through photosynthesis every day on land worldwide (Vitousek et al., 1986).

The human harvest of NPP from the oceans, which are much less productive than land despite their far greater extent, is a comparatively modest 2% (Vitousek et al., 1986). In ocean systems, NPP is produced by highly dispersed phytoplankton and algae, which are not likely soon to become an important part of the human diet. Rather, we will continue depending on fishes, which feed mostly near the tops of marine food webs and conserve only a small fraction of the NPP. Human impacts on oceanic ecosystems, however, are not negligible. The increasing pollution of coastal areas and accelerating depletion — even commercial extinction — of many important fish stocks is putting that critical food source in jeopardy (World Resources Institute et al., 1992; Weber, 1993).

The bulk of the human impact on global NPP is on land, and is indirect in nature. Some 30% of terrestrial NPP (including that consumed directly by societies) is diverted into human-directed ecosystems composed of different organisms than would otherwise be present. Obvious examples are croplands, tree farms, heavily managed forests, and converted pastures, as well as urban plantings, parks, etc. Terrestrial NPP is further affected by human activities and land-use changes that prevent or reduce the amount of photosynthesis that would otherwise occur in many areas. Reduction occurs through degradation of natural or agricultural ecosystems, or by conversion of ecosystems into less productive ones, such as forests to cropland or pasture. And no NPP is produced on land that is paved or built over. It has been estimated that, since

1950, about 13% of potential terrestrial NPP has been lost in various ways, bringing the total human impact to over 40% on land, and about 25% globally (Vitousek *et al.*, 1986).

The diversion of such a large portion of the world's NPP into the human account goes far to explain the disappearance of biodiversity — a loss that is only beginning to be properly accounted for and appreciated (Ehrlich and Ehrlich, 1981; Wilson, 1992). It may also have been partially captured in the assessment by UNEP of the degradation of the world's productive lands (Oldeman *et al.*, 1990).

VII. MEASURING IMPACTS

Human beings today can be seen as a force that is mobilizing resources on a scale comparable to, or in some cases exceeding, natural cycles. Civilization is now also threatening to destabilize major elements of the natural life-support system through such changes as large-scale deforestation (Myers, 1989; World Resources Institute *et al.*, 1992), depletion of the stratospheric ozone shield, and the build-up of greenhouse gases in the atmosphere (Houghton *et al.*, 1990; Ehrlich and Ehrlich, 1991). A civilization that allows the degradation of its essential resource base and destabilization of global climate and biogeochemical processes is very far from sustainable.

Two decades ago, Paul Ehrlich and John Holdren devised a simple equation to describe in very simplified terms the multiplicative effects of numbers of people and their use of resources in generating environmental impacts (Ehrlich and Holdren, 1971; Ehrlich and Ehrlich, 1991). The equation (more accurately, an identity) is:

$$I = P \times A \times T$$

where I represents the total environmental impact (of a society); P is the population; A is the affluence (or consumption) of each individual on average; and T is the technology used to supply each unit of consumption.

Although reasonably accurate statistics exist for population sizes, no useful measure exists for total consumption per person or for technology. The latter two are especially difficult to untangle; indeed the three factors are by no means independent of each other. A rough idea of the relationship can be captured by substituting per-capita energy use for A $\times$ T (AT). Considering the central role that energy use plays in generating most environmental impacts, this is a useful approximation, if greatly oversimplified, of the range and scope of human impacts.

By this measure, citizens of industrialized nations can be seen on average to have seven times the impact per person of people in the developing world. At the extremes, the average US citizen (whose energy use is 50% to 100% higher than that of the average European or Japanese) has six times the impact of a person in Mexico and about 30 times that of someone in a very poor nation such as Bangladesh or Chad (Ehrlich and Ehrlich, 1991). Thus, although industrialized nations have only about a fifth of the world's people (1.2 billion), and their populations are growing slowly if at all, they control and consume half to four-fifths of most resources, including food and agricultural inputs (World Bank, 1992; World Resources Institute *et al.*, 1994). Consequently, they are causing far more than their share of damage to the Earth's life-support systems.

People in industrialized nations sometimes claim that the "population problem" is a problem of developing countries, because populations in rich nations are growing so slowly. But spokespeople of the developing world often assert that the problem is not too many poor people, but overconsumption and overexploitation by the rich (e.g., Makhijani, 1992). As the $I = PAT$ identity indicates, both views are partly right. Yet, overconsumption in rich societies is driven at least in part by overpopulation. If only half as many people were consuming at the same rate, they certainly would be doing far less damage.

Meanwhile, the poor aspire to consume as the rich do. In societies that are rapidly industrializing, population growth drives up consumption rates and pressures on resources. Many of these "middle-income" societies have prospering industrial sectors and a widening income gap, with the lower end often occupied by the rural poor.

In the poorest societies, where high birth rates still prevail, population growth often intensifies pressure on the land, accelerating land degradation and deepening rural poverty (Kates and Haarmann,1992; World Bank, 1992). Sub-Saharan Africa includes many of the world's poorest nations, where population growth rates of 3% per year or more have clearly contributed to environmental degradation and hindered development efforts. The region has suffered a decline in both food production and GNP per capita of about 20% over the last 25 years (Durning, 1989; Darnton, 1994). Smaller declines in per capita food production have also occurred in Latin America and some other poor nations.

VIII. NUTRITIONAL SECURITY

Most people would agree that a nutritionally secure society is one able to provide all its people with diets sufficient to sustain work and other normal daily activities (Ehrlich *et al.*, 1993). This security might be achieved through domestic food production or through an ability to purchase or trade for foods produced in surplus elsewhere. A secure food system includes buffers against both poor harvests and problems in obtaining food through trade. Most countries have such buffers in some degree today. Food has been traded on the global market for several decades, allowing shortages in most areas to be compensated by surpluses elsewhere through trade. For very poor nations that cannot afford to buy food on the world market, the World Food Programme and various private agencies provide emergency supplies in times of famine.

Naturally, most nations see it in their own interest to maintain some degree of self-sufficiency in food production, no matter how well integrated the world food market and distribution system may be. But when global food production falls significantly below consumption levels, as it did in 1988, there is no guarantee that sufficient food will be available at reasonable prices. Moreover, at such times, the World Food Programme often finds it harder than usual to collect food donations for poor countries.

Perhaps the global food distribution system has given people in rich countries an undeserved sense of security. A widespread impression prevails that the green revolution has more or less permanently solved the problem of feeding the growing population and that famine has been largely banished, except for local famines traceable to political conflicts (Swaminathan and Sinha, 1986). It is often claimed that the persisting widespread chronic undernourishment is due simply to maldistribution of otherwise abundant food supplies and that better distribution would solve the hunger problem (Waggoner, 1994).

Complacency about the security and abundance of the world food supply, however, is not supported by careful analysis, especially if the environmental dimensions of the agricultural enterprise are carefully considered. It is true that outright starvation today is usually a result of food distribution failures rooted in politics, as are the tragic situations in Somalia and Sudan, and earlier in Ethiopia (Dreze and Sen, 1991). But, although these acute cases attract much public attention, they usually affect only a few hundred thousand people at a time. They are thus a tiny tip of the iceberg of widepread hunger in developing nations, whose causes are less straightforward. The acute starvation in poor nations in turmoil, moreover, has usually been precipitated in an already vulnerable, poorly nourished population.

Estimates of the actual level of chronic hunger in developing nations in the 1990s vary among international agencies, but the consensus ranges between 750 million and over a billion, that is, between 15% and 20% of the world's population (e.g., FAO, 1992; World Bank, 1992). At least half of the 14 million children who die each year of preventable causes in poor countries succumb because they were malnourished (UNICEF, 1992). The hungry millions are concentrated in South Asia and sub-Saharan Africa, where they comprise up to half of the populations. Smaller fractions of populations in many other developing regions are also undernourished.

In the strictest sense, the widespread chronic food shortages in many developing regions can be assigned largely to maldistribution resulting from poverty and related economic factors, including inequities in the world trade system. Too often also, developing nations have neglected their agricultural sectors — "taxed" them, in economic parlance — in favor of industrial development. A World Bank study has shown the shocking degree to which industry and the urban sector have been subsidized directly and indirectly (through price interventions, import and export taxes, and by manipulating exchange rates) by agriculture in many developing nations (Schiff and Valdes, 1992). These burdens have far outweighed minor subsidies that governments provide to farmers for seeds and inputs. The result has been to retard rural development and perpetuate poverty. Most important, no benefits were provided to national economies; rather, those countries with the highest taxes on the farm sector generally experienced slower economic growth.

One result of rural poverty has been massive migration from countryside to cities, often overwhelming the cities' ability to accommodate the influx. Sometimes the rural–urban shift has been a result of green revolution success, when better-off farmers have stopped hiring landless workers and switched to mechanized farming. Poorer farmers who own land may lose out in the competition, unable to afford the high-yield seeds and inputs (Dahlberg, 1979; World Bank, 1992).

Many migrants travel further; the United Nations has estimated that more than 100 million people lived outside their nation of origin by the early 1990s (United Nations Population Fund, 1993). Many international migrants have been economic refugees simply seeking a better life (most of Europe's guest workers fit this category). Many have been political refugees, fleeing wars and upheaval in nations as disparate as Vietnam, Ethiopia, Bosnia, and El Salvador. But more and more migrants today are in reality environmental refugees fleeing from damaged and exhausted lands (Jacobson, 1988). And when poverty and resource pressures lead to conflict, the distinction between political and environmental refugees becomes increasingly blurred (Homer-Dixon et al., 1993). Obvious and poignant cases in point are Haiti and Rwanda.

IX. THE RICH-POOR GAP, GLOBAL CHANGE, AND AGRICULTURE

Until the recent emergence of environmental problems on a global scale, the differences in impact among richer or poorer societies were not considered an issue; most problems were local, or at most national, and solutions were sought locally. Since the 1970s, however, environmental problems have been seen as increasingly international: acid precipitation, depletion of the stratospheric ozone shield by CFCs, and global warming. As a consequence, the world community has been forced to address these problems through international cooperation.

Even though the populations of developing nations are collectively larger and growing much faster (at an average annual rate of about 2.2%, excluding China, in 1994), their aggregate contribution to human-caused global change is still smaller than that of the rich — a circumstance that has led to much controversy in discussing solutions to global environmental problems. But the share of developing nations' responsibility for these problems, especially global warming, is rising rapidly, and the potential for far greater contributions in the near future is huge, given substantial growth in both populations and economies (Ehrlich and Ehrlich, 1991).

Indeed, a recent report indicated that the developing world's annual contributions of carbon dioxide emissions to global warming now exceed those of the industrialized nations and are rising faster. The carbon dioxide build-up is mostly the result of combustion of fossil fuels, although deforestation and land clearing also make contributions. Two other important human-caused greenhouse gas emissions, methane and nitrous oxide, may also be largely from developing countries, although the sources and sinks of those gases are still too poorly known to assign responsibility accurately (Ehrlich, 1990; Houghton et al., 1990).

Perhaps more important, emissions of the latter two gases are significantly associated with agriculture. Major sources of methane emissions include rice paddies and flatulence of cattle, as well as land clearing, landfills, wetlands, and leaks from natural gas pipelines. Nitrous oxide emissions are strongly linked to use of nitrogen fertilizers as well as land use changes. Both are many times more potent greenhouse gases than carbon dioxide, and methane concentrations until recently have been building even more rapidly in the atmosphere, although methane has a relatively short residence time there.

Although agricultural activities are very important in generating environmental problems from pesticide pollution to loss of biodiversity, land degradation, and greenhouse gas emissions, the consequences of many environmental problems include impacts on agriculture (Ehrlich and Ehrlich, 1991). Some, such as air pollutants and acid precipitation originate mostly off the farm, but are known to depress crop yields. Similarly, the increased flux of ultraviolet light reaching the surface, caused by depletion of the ozone layer by CFCs, has damaging effects on many crops.

But most worrisome is the potential for disruption of agriculture worldwide by unpredictable and possibly severe changes in climate (Houghton et al., 1990; Parry, 1990). Agriculture is always closely adapted to local climate; any change, even a potentially beneficial one, will be disruptive in the short run until farmers adjust to it. Any climate change caused by global warming is unlikely to be a simple shift to a new regime; rather local climate in most places will become a moving target, with the past no predictor of the future. The effect on world food production is even less predictable, but common sense suggests that a reduction in harvests is more likely than a sharp increase over the medium term (Daily and Ehrlich, 1990).

X. THE OUTLOOK

Given the trends in the condition of our natural capital, the green revolution's loss of steam, and the possibility that mounting environmental problems will increasingly constrain food production, Lester Brown, of Worldwatch Institute, several years ago warned that the global grain harvest in the 1990s was unlikely to expand as fast as the population (Brown, 1988). So far, even though the 1990s have brought a slight slackening of population growth (Haub, 1992, 1994), history has not contradicted that prognosis. Fig. 2 graphically compares the increases in global grain production between 1950 and 1993 and the changes in production per capita. A near tripling of production in absolute terms has not been matched by an increase per capita, even though until 1985 the latter did rise more or less steadily, because of population growth. Much of the per-capita increase, however, was used for feed as the more affluent populations increased their meat consumption.

Although many opportunities for increasing food production have been foreclosed in the last half-century, some others remain open. Setting aside the urgency of reducing population growth — which is most needed in the most nutritionally precarious societies, and without which all efforts to expand food supplies would ultimately fail — more attention is needed for other solutions.

One relatively easy way to increase food supplies is simply to reduce the amount that is lost to pests and spoilage after harvest (Ehrlich *et al.*, 1993). Estimates of losses on a worldwide basis range from 25% to 40% (Greely, 1991). The greatest losses occur in developing nations where storage and transport facilities are often inadequate or entirely lacking and where climates are also conducive to rapid spoilage. Saving only a quarter of the estimated losses would make available enough food to support hundreds of millions of people.

Many observers correctly point out that a very large fraction of world grain production — currently about a third — is used to feed livestock. This is a

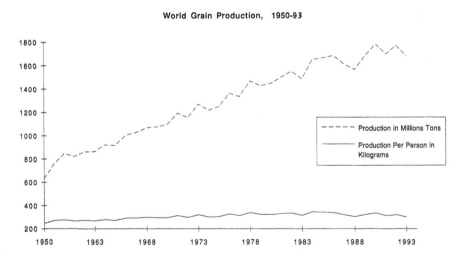

Fig. 2.   World grain production 1950–93. Data from Brown *et al.* (1994b).

relatively inefficient use of food resources since the amount of calories in the feed is several times more than the calories available in the meat, eggs, or milk produced — a range of two to seven or more times, depending on the product (Pimentel and Hall, 1989). The reason that nobody seems to have noticed that per-capita food production declined after 1984 is that per-capita consumption of meat and other animal products has declined (apparently due to lower demand, not to higher prices), and the share of the world's grain used for feed has fallen from nearly 40% in the early 1980s to about a third in the 1990s. Today, although many people in rich countries are reducing their animal product consumption for health reasons, people in developing nations who have reached income levels at which they can afford these luxury foods are increasing their consumption, and many development observers expect this trend to continue. Nevertheless, the worldwide slackening of economic growth that has also occurred since the late 1980s probably also played a part in lowering demand for animal products in many regions (Brown et al., 1994).

Of course, the large share of grain used as feed represents a huge potential source of human food — assuming that all the people who eat animal products could be persuaded to give them up, and assuming that more palatable grains could be grown in place of feed crops. Some agricultural specialists are complacent that many billions more people could be fed in the future without wrecking environmental underpinnings largely by giving up livestock feeding (Waggoner, 1994). Such a change would imply great faith in the feasibility of eliminating the gaps between rich and poor in many dimensions, of which food is only one, albeit an important one.

Although the green revolution has already wrought most of its miracles, it may still offer some further increases in yields, especially if applied to crops that until now have been overlooked, especially crops of the humid tropics such as cassava. Many potentially useful traditional food crops have been neglected or forgotten, sometimes under the impact of colonialism. Seeking them out, improving them genetically, and reintroducing them could provide a needed boost to food production in many developing regions.

Even though most of the world's agriculturally suitable land is already in use, in some areas good land remains that is underutilized. This is especially the case in Latin America, where land ownership practices often keep good land underused while intensive subsistence farming is relegated to poorer, more marginal land. Too often, though, assumptions are made that "unused" land, such as that beneath the forests of the Amazon basin or Central Africa, are a potential source of agricultural abundance (e.g., FAO and IIASA, 1982). Decisions about land conversions must be based on careful assessments of the best use of the land, not based on narrow, short-term economic considerations (Goodland et al., 1984; Fearnside, 1987).

The example of China's surge in food production in the 1980s when peasants were given the right to profit from their labor suggests that economic arrangements can have enormous impacts on food production, either hindering or stimulating it. The downside of China's production surge, however, was that much of it led to an acceleration in land degradation, thus creating a temporary increase in harvests at the cost of future production. The explosive Chinese economy in recent years has stimulated a building boom that is rapidly subtracting prime land from the cropland base (Brown, 1994). At the same time, prospering Chinese people are developing a taste for animal products that is likely to increase demand for grain far faster than the population is growing. One Chinese scientist gloomily noted that if demand continued to rise while losses of cropland proceeded at current rates, and yields were not significantly increased, China would need to import more grain in 2030 than the United States produces to fill the gap (Brown, 1994).

The former Soviet Union, already handicapped by a difficult and undependable climate in much of its territory, compounded its problems by making costly mistakes in land management and operating state farms in highly inefficient ways. Converting to a more efficient agricultural system clearly will take some time, but there is hope that significantly more food can be produced in that region than now. Important lessons are to be learned in both China and Russia, of course, with regard to managing fragile lands.

Economic adjustments in many developing nations might well make agricultural sectors more productive, while at the same time improving the lot of the poor and reducing environmental impacts (Feyerabend, 1994; Dreze and Sen, 1990–91). How much such changes might increase food production is unknown and will no doubt vary greatly from place to place (Vosti *et al.*, 1991). However, it is very clear that even modest assistance to very poor individuals and communities can work apparent miracles in stimulating productivity and often in preserving environmental resources as well (Borrini, 1990; Norse, 1992; Dasgupta, 1993).

Finally, what is most needed is a cold, hard look at the green revolution in an ecological context. The next revolution indeed may well be the ecological one (if one can judge from the outpouring of literature on the subject). Although it has scarcely begun to be implemented on a significant scale (except in areas where traditional farming methods have survived), initial results and findings are promising (Altieri,1983; Lowrance *et al.*, 1984; Dover and Talbot, 1987; Carroll *et al.*, 1990; Soule and Piper, 1992). Not only is there considerable interest in finding more ecologically sensible ways of farming in the tropics, but there is also research aimed at finding more sustainable approaches to farming in temperate zones (National Research Council, 1989; Edwards *et al.*, 1990).

In many situations, new farming systems will need to be developed that may be radically different from those viewed as conventional now. In particular, the use of mixed crops, more frequent crop rotation, and other apparently old-fashioned methods, when combined with higher-yielding crops and sophisticated fertilizing methods, may turn out to be surprisingly effective. But it is clear that the new approach will not be a single bundle of technology to be universally applied, regardless of climate or soil characteristics. Rather, the best results might be obtained by working with the soils and trying out a variety of crop combinations and strategies for pest and weed management to find the best system (or set of systems to be varied through time, thus confounding pests) in each locality (Pimentel *et al.*, 1992).

Pest management itself is an area where the need for new approaches is clear, given the rates at which pests develop resistance to pesticides and the human health and environmental problems they cause (Ehrlich and Ehrlich, 1991). The increasing interest in and practice of "integrated pest management" is an encouraging change for the better (Holl *et al.*, 1990; Horn, 1988).

Many of the "new" ideas of ecological agriculture would be very familiar to traditional farmers in developing regions, and indeed, many have been adapted from that source. But of course, traditional farming was developed in an experimental fashion over centuries as farmers learned what worked best and most sustainably in their particular ecosystems. Modern agronomists can learn much from the traditional farmers, and the farmers could benefit from the fruits of science in improving their crop yields, protecting resources, managing pests and weeds, and increasing food production for their society — if given the chance. Clearly, increasing food production while creating a sustainable system will require a multidisciplinary approach, just as will solving some of the other major problems faced by the world community. We have little choice but to try.

The human enterprise cannot be put on a truly sustainable basis until it is recognized that humanity is now living beyond its means and that a path towards a viable future must be carved out and agreed on. The next century or two will not be easy ones, when the principal task will be to manage the transitional overshoot phase, while populations and resource use exceed regional carrying capacities and stresses on life-support systems continue to mount. Humanely reversing the population explosion itself will require many decades to accomplish; so will reducing the material throughput of over-industrialized societies, kicking the fossil fuel habit and curbing greenhouse gas emissions, detoxifying and restoring damaged and decimated ecosystems, developing sustainable, productive agricultural systems, providing the basic necessities of life for the poor in every nation, and somehow avoiding serious international warfare in the process.

Without some clear vision of an attractive, attainable shared future, these difficult decisions are not likely to be made. Each society needs to evaluate its own territory's carrying capacity, especially its nutritional security, and including its options for trade with other societies. Then it should decide on its preferred living standard and lifestyle and reach a consensus on a population size that fits those parameters, allowing an ample margin for error. Appropriate policies should be adopted for easing the transition.

Of course, all these decisions and policies will demand an unprecedented level of cooperation among nations. Although some progress has been made in international cooperation at the political level, even as economies are increasingly tied together in a global network, much more is needed. We can all be grateful that the end of the cold war presents an opportunity to create a new cooperative world structure — an opportunity we cannot afford to miss.

REFERENCES

Altieri, M. (1983). "Agroecology: The Scientific Basis of Alternative Agriculture", Division of Biological Control, University of California, Berkeley.

Borrini, G. (1990). "Lessons learned in Community-Based Environmental Management". International Course for Primary Health Care Managers at District Level in Developing Countries. Instituto Superiore Di Sanita, Rome.

Brown, L. R. (1988). The changing world food prospect: The nineties and beyond, *Worldwatch Report 85*, Worldwatch Institute, Washington DC.

Brown, L. R. (1991). Fertilizer engine losing steam. *World Watch* 4(5), 32–33.

Brown, L. R. (1994). Who will feed China? *World Watch* 7(5), 10–17.

Brown, L. R. and Wolf, E. C. (1984). Soil erosion: quiet crisis in the world economy, *Worldwatch Paper 60*, Worldwatch Institute, Washington DC.

Brown, L. R., Durning, A., French, H., Lenssen, N., Lowe, M., Misch, A., Postel, S., Renner, M., Starke, L., Weber, P. and Young, S. (1994a). "State of the World 1994", W. W. Norton, New York.

Brown, L. R., Kane, H. and Roodman, D. M. (1994b). "Vital Signs 1994", p. 27. Worldwatch Institute, Washington, DC.

Bugbee, B. and Monje, O. (1992). The limits of crop productivity. *BioScience* 42, 494–502.

Carroll, C., Vandermeer, J., Rosset, P. (eds) (1990). "Agroecology". McGraw-Hill, New York.

Cohen, J., Williams, J., Plunkett, D. and Shands, H. (1991). Ex situ conservation of plant genetic resources: global development and environmental concerns. *Science* 253, 866–872.

Dahlberg, K. (1979). "Beyond the Green Revolution," Plenum, New York.

Daily, G. C. and Ehrlich, P. R. (1990). An exploratory model of the impact of rapid climate change on the world food situation. *Proceedings of the Royal Society of London* 241, 232–244.

Daily, G. C. and Ehrlich, P. R. (1992). Population, sustainability, and Earth's carrying capacity, *BioScience* 42, 761–771.

Daly, H. E. and Cobb, J. E., Jr (1989). "For the Common Good", Beacon Press, Boston.

Darnton, J. (1994). "Lost decade" drains Africa's vitality. *New York Times*, June 19, p. 1.

Dasgupta, P. (1993). Fertility and resources: the household as a reproductive unit. *In* "An Inquiry into Well-Being and Destitution" (P. Dasgupta, ed.), pp. 343–370. Oxford University Press, Oxford.

Dover, M. and Talbot, L. (1987). "To Feed the Earth: Agro-Ecology for Sustainable Development", World Resources Institute, Washington, DC.

Dreze, J. and Sen, A. (eds) (1990–1991). "The Political Economy of Hunger." Vol. I: "Entitlement and Well-being" (1990). Vol. II: "Famine Prevention" (1990). Vol. III: "Endemic Hunger" (1991). Oxford University Press, Oxford.

Durning, A. B. (1989) Poverty and the environment: reversing the downward spiral. *Worldwatch Paper 92*, Worldwatch Institute, Washington DC (November).

Edwards, C., Lal, R., Madden, P., Miller, R. and House, G., (eds) (1990). "Sustainable Agricultural Systems". Soil and Water Conservation Society, Ankeny, Iowa.

Ehrlich, A. H. (1990). Agricultural contributions to global warming. In "Global Warming: The Greenpeace Report" (J. Leggett, ed.), pp. 400–420. Oxford University Press, Oxford and New York.

Ehrlich, A. H. (1994). Building a sustainable food system. In "The World at the Crossroads: Towards a Sustainable, Equitable and Liveable World" (P. B. Smith, S. E. Okoye, J. de Wilde and P. Deshingkar, eds), pp. 21–38. Earthscan, London.

Ehrlich, P. R. (1993). Biodiversity and ecosystem function: need we know more? Foreword in "Biodiversity and Ecosystem Function" (E.-D. Schulze and H. Mooney, eds), Springer-Verlag, Berlin.

Ehrlich, P. R. and Ehrlich, A. H. (1981). "Extinction: The Causes and Consequences of the Disappearance of Species", Random House, New York.

Ehrlich, P. R. and Ehrlich, A. H. (1991). "Healing the Planet", Addison-Wesley, Reading, MA.

Ehrlich, P. R. and Ehrlich, A. H. (1992). The value of biodiversity. Ambio 21, 219–226.

Ehrlich, P. R. and Holdren, J. P. (1971). The impact of population growth, Science 171, 1212–1217.

Ehrlich, P. R. and Mooney, H. (1983). Extinction, substitution, and ecosystem services, BioScience 33, 248–254.

Ehrlich, P. R. and Wilson, E. O. (1991). Biodiversity studies: Science and policy. Science 253, 758–762.

Ehrlich, P. R., Ehrlich, A. H. and Daily, G. C. (1993). Nutritional security, population, and food. Population and Development Review 19(2), 1–32.

Falkenmark, M. and Widstrand, C. (1992). Population and water resources: a delicate balance. Population Bulletin 47(3), 1–36.

FAO (Food and Agriculture Organization) (1992). "World Food Supplies and Prevalence of Chronic Undernutrition in Developing Regions as Assessed in 1992." FAO, Rome.

FAO (Food and Agriculture Organization) and IIASA (International Institute for Applied Systems Analysis) (1982). "Potential Population Supporting Capacities of Lands in the Developing World". FAO, Rome.

Fearnside, P. (1987). Rethinking continuous cultivation in Amazonia. BioScience 37, 209–214.

Feyerabend, G. B. (1994). People's empowerment: a condition for a sustainable, liveable, and equitable society? In "The World at the Crossroads: Towards a Sustainable, Equitable and Liveable World" (P. B. Smith, S. E. Okoye, J. de Wilde and P. Deshingkar, eds), pp. 125–136. Earthscan, London.

Gasser, C. S. and Fraley, R. T. (1989). Genetically engineering plants for crop improvement. Science 244, 1293–1299.

Gleick, P. (1993). "Water in Crisis", Oxford University Press, New York.

Goodland, R., Watson, C. and Ledec, G. (1984). "Environmental Management in Tropical Agriculture". Westview, Boulder, CO.

Greely, M. (1991). Postharvest losses — the real picture. International Agricultural Development, Sept./Oct., pp. 9–11.

Grigg, D. (1993). "The World Food Problem", Blackwell, Oxford.

Haub, C. (1992). China's fertility drop lowers world growth rate. Population Today 21(6) (June). Population Reference Bureau, Washington, DC.

Haub, C. (1994). World growth rate slows, but numbers build up, Population Today

22(11) (November). Population Reference Bureau, Washington, DC.

Haub, C. and Yanagishita, M. (1994). *1994 World Population Data Sheet*, Population Reference Bureau, Washington DC. (This is the source for all current population figures unless otherwise noted.)

Hillel, D. (1991). "Out of the Earth: Civilization and the Life of the Soil". The Free Press (Macmillan), New York.

Holdren, J. P. (1990). Energy in transition, *Scientific American* **26**(3), 156–163.

Holl, K., Daily, G. and Ehrlich, P. (1990). Integrated pest management in Latin America. *Environmental Conservation* **17**, 341–350.

Homer-Dixon, T., Boutwell, J. and Rathjens, G. (1993). Environmental change and violent conflict, *Scientific American* **268**(2), 16–25.

Horn, D. (1988). "Ecological Approaches to Pest Management", Elsevier, London.

Houghton, J., Jenkins, G. and Ephraums, J. (eds) (1990). "Climate Change: The IPCC Scientific Assessment". Intergovernmental Panel on Climate Change (IPCC), Cambridge University Press, New York and Cambridge.

Hoyt, E. (1988). "Conserving the Wild Relatives of Crops", International Board for Plant Genetic Resources, International Union for the Conservation of Nature and Natural Resources (IUCN), and Worldwide Fund for Nature (WWF), Rome and Gland.

IRRI (International Rice Research Institute) (1992). Yield stagnation, yield decline, and the yield frontier of irrigated rice. "Program Report to the Board Program Committee". 21–23 September. Mimeo.

Jacobson, J. (1988). Environmental refugees: a yardstick of habitability. *Worldwatch Paper 86*, Worldwatch Institute, Washington, D.C.

Kates, R. W.and Haarmann, V. (1992). Where the poor live. *Environment* **34**(4), 4–11, 25–28.

Keyfitz, N. (1991). Population and development within the ecosphere: one view of the literature. *Population Index* **57**(1), 5–22.

Lee, R., Arthur, W., Kelley, A., Rodgers, G. and Srinivasan, T. (eds) (1988). "Population, Food, and Rural Development". Clarendon Press, Oxford.

Lowrance, R., Stinner, B. and House, G. (1984). "Agricultural Ecosystems: Unifying Concepts". Wiley/Interscience, New York.

Maass, J. and Garcia-Oliva, F. (1992). Erosion de suelos y conservation biologica en Mexico y Centroamerica. *In* "Conservation y Manejo de Recursos Naturales en America Latina" (R. Dirzo, D. Pinera and M. Kalin-Arroyo eds), Red Latinoamericana de Botanico, Santiago, Chile.

Makhijani, A. (1992). "From Global Capitalism to Economic Justice", Apex Press, New York.

Matthews, E. (1983). Global vegetation and land use: new high-resolution data bases for climate studies. *Journal of Climate and Applied Meteorology* **22**, 474–487.

McNicoll, G. (1992). The United Nations' long-range population projections. *Population and Development Review* **18**, 333–340.

Myers, N. (1989). "Deforestation Rates in Tropical Forests and their Climatic Implications". Friends of the Earth, London.

National Research Council, Board on Agriculture (1991). "Managing Global Genetic Resources". National Academy Press, Washington DC.

National Research Council, Committee on the Role of Alternative Farming Methods in Modern Production Agriculture (1989). "Alternative Agriculture". National Academy Press, Washington DC.

Naylor, R. (1994). Herbicide use in Asian rice production. *World Development* **22**, 55–70.

Norse, D. (1992). A new strategy for feeding a crowded planet. *Environment* **34**(5), 6–11, 32–39.

Oldeman, L., Van Engelen, V. and Pulles, G. (1990). The extent of human-induced soil degradation. Annex 5 of Oldeman, L. *et al.*, "World Map of the Status of Human-Induced Soil Degradation: An Explanatory Note." rev. 2nd edition. International Soil Reference and Information Centre (ISRIC), Wageningen, The Netherlands.

Paddock, W. (1990). Our last chance to win the war on hunger. *Population and Environment* 11(3).

Paddock, W. (1992). Our last chance to win the war on hunger. *Advances in Plant Pathology* 8, 197–222.

Parry, M. (1990). "Climate Change and World Agriculture". Earthscan, London.

Pimentel, D. (ed.) (1993). "World Soil Erosion and Conservation". Cambridge University Press, Cambridge.

Pimentel, D. and Hall, C. (eds) (1989). "Food and Natural Resources". Academic Press, San Diego.

Pimentel, D., Stachow, U., Takacs, D., Brubaker, H., Dumas, A., Meany, J., O'Neal, J., Onsi, D. and Corzilius, D. (1992). Conserving biological diversity in agricultural/forestry systems. *BioScience* 42(5), 354–362.

Pluncknett, D., Smith, N., Williams, J. and Anishetty, N. (1987). "Gene Banks and the World's Food". Princeton University Press, Princeton, NJ.

Postel, S. (1990). Water for agriculture: facing the limits, *Worldwatch Paper 93*, Worldwatch Institute, Washington DC.

Sanchez, P. (1976). "Properties and Management of Soils in the Tropics". Wiley-Interscience, New York.

Schiff, M. and Valdes, A. (1992). "The Plundering of Agriculture in Developing Countries". World Bank, Washington DC.

Soule, J. and Piper, J. (1992). "Farming in Nature's Image". Island Press, Washington DC.

Stevens, W. K. (1994). Feeding a booming population without destroying the planet. *New York Times*, April 5 (science section).

Swaminathan, M. and Sinha, S. (eds) (1986). "Global Aspects of Food Production". Tycooly International, Riverton.

Turner, B. II, Clark, W., Kates, R., Richards, J., Mathews, J. and Meyer, W. (eds) (1990). "The Earth as Transformed by Human Action". Cambridge University Press, Cambridge.

Tyler, P. E. (1993). China lacks water to meet its mighty thirst, *New York Times*, Nov. 7.

UNICEF (United Nations Children's Fund) (1992). "State of the World's Children, 1992". United Nations, New York.

United Nations Population Division (1991), "Long-Range World Population Projections", ST/SEA/SER.A/125, UN, New York.

United Nations Population Fund (UNFPA) (1992). "State of the World Population, 1992". United Nations, New York.

United States Department of Agriculture (USDA). Foreign Agriculture Service (1994). *World Agricultural Production*, WAP 5-94 (May). Washington DC.

Vitousek, P. M., Ehrlich, P. R., Ehrlich, A. H. and Matson, P. A. (1986). Human appropriation of the products of photosynthesis. *BioScience* 36, 368–373.

Vosti, S., Reardon, T. and von Urff, W. (eds) (1991). "Agricultural Sustainability, Growth, and Poverty Alleviation: Issues and Policies". (Proceedings of conference 23–27 September 1991, Feldaring, Germany. Available from Deutsche Stiftung fur Internationale Entwicklung (DSE), Zentralstelle fur Ernahrung und Landwirtschaft, Wielinger Str. 52, D-8133 Feldafing, Federal Republic of Germany).

Waggoner, P. (1994). "How Much Land Can Ten Billion People Spare for Nature?" Council for Agricultural Science and Technology, Ames, Iowa.

Walsh, J. (1991). Preserving the options: food production and sustainability. *Issues in Agriculture No. 2*, Consultative Group on International Agriculture Research (October).

Weber, P. (1993). Reversing the decline of the oceans. *Worldwatch Paper 116*, Worldwatch Institute, Washington DC.

Wilson, E. O. (1989). Threats to biodiversity. *Scientific American* **261**(3), 108–116.

Wilson, E. O. (1992). "The Diversity of Life". Harvard University Press, Cambridge MA.

Wilson, E. O. and Peter, F. (1988). "Biodiversity", National Academy Press, Washington DC.

World Bank (1992). "World Development Report 1992: Development and the Environment". Oxford University Press, Oxford.

World Commission on Environment and Development (1987). "Our Common Future". Oxford University Press, London and New York.

World Resources Institute, United Nations Environment Programme, and United Nations Development Program (1992). "World Resources 1992–93". Chapters 7 and 8, Oxford University Press, New York.

World Resources Institute, United Nations Environment Programme, and United Nations Development Program (1994). "World Resources 1994–95". Chapter 6. Oxford University Press, New York.

Wrigley, G. (1982). "Tropical Agriculture: The Development of Production". Longman, New York.

Plant Virus Infection: Another Point of View*

G. A. DE ZOETEN

Department of Botany and Plant Pathology, Michigan State University, East Lansing, Michigan 48824 USA

I.   INTRODUCTION

Plant virus infection is generally believed to proceed in the following manner: (1) virus enters the plant cell through a wound (Matthews, 1981) made either by an abrasive or a vector; (2) replication starts after virus uncoating is effected *within* the cell; and (3) systemic invasion is mediated by transport of progeny virus that initiate the infection in cells either distant from or in close proximity to the initial infection, where uncoating and replication are then repeated.

---

* Editor's note: Periodically, *Advances in Botanical Research* will present articles that are essentially editorial in theme. The opinions expressed are those of the authors.

Advances in Botanical Research Vol. 21
incorporating Advances in Plant Pathology
ISBN 0–12–005921–5

Critical reassessment of this classic scheme of plant virus infection in the light of current knowledge provided the basis for the following hypothesis to be considered here: (1) A cascade of virus-destabilizing events, initiated at the cell wall–virus interface (cuticle) and terminated after the genomic RNA passes the plasmalemma, comprises uncoating in the infection cycle. Similar destabilization is not repeated during subsequent invasion of the host. (2) Replication is initiated utilizing a host membrane system (replication complex) as the backbone for genome transcription. Subsequent coat protein production and particle assembly removes most progeny RNA from the infection cycle since the necessary cell surface environment for initiation of destabilization and uncoating of progeny virions is absent in the symplast. (3) Cell-to-cell and long distance infection spread (invasion) are two different processes and are mediated by transport of the replication complex which, through membrane fusion, initiates replication in compartments of the symplast hitherto uninfected.

II. THE PLANT AS A MEDIUM FOR VIRUS REPLICATION

The phenomena that set the biology of plants apart from that of animals, and the viral interactions that alter this biology, have to be addressed first, since understanding these phenomena is crucial for the discussion that follows.

Over the past twenty years (see reviews of Gunning and Robards, 1976; Robards and Lucas, 1990; and Lucas *et al.*, 1993) innovative and intensive experimentation has led to the understanding that plasmodesmata provide for crucial venues of molecular and particulate communication between the compartments of the symplast on either side of the intervening cell walls.

In essence, the plant symplast is a continuum (cytosol) that is divided into functional domains by plasmodesmata that can regulate simple diffusion processes, complex macromolecular trafficking and particulate exchanges between the domains they delimit (Lucas *et al.*, 1993). They state (p. 436):

We suggest that the evolution of these plasmodesmal properties has established the unique condition whereby plants function as supracellular, rather than multicellular organisms. As such, the dynamics of the plant body, including cell differentiation, tissue formation, organogenesis, and specialized physiological function(s), would be subject to plasmodesmal regulation.

The size exclusion limit (SEL) for trafficking and communication exhibited by different plasmodesmata in different tissues (domains) is extremely varied and may well determine the differentiation of domains themselves. The concepts developed in regard to plasmodesmatal evolution and formation should have momentous effects on our understanding of virus infection as do the concepts and emerging understanding of viral movement and its effect on plasmodesmatal trafficking. In this regard, the alliance between those trying to understand plasmodesmatal function and those studying viral movement in plants promises to substantially advance both fields of endeavor (Lapidot *et al.*, 1993;

Noueiry *et al.*, 1994).

Even superficial study of the seminal papers that reconceptualize and redefine the evolutionary role and the function of plasmodesmata leads to the realization that the current global view of plant virus infection is at odds with current concepts in plant biology in regard to plasmodesmata and their function in a supracellular organism and, indeed, with the concept of supracellular organisms itself. First, the animal virus model of infection, viral replication — virus release followed by subsequent infection of uninfected cells in multicellular organisms — may not be the way infection spread occurs in supracellular organisms like plants, where viral movement proteins open up domains for subsequent infection which is aided by the fact that the cytosol in effect is a continuum throughout the plant. Secondly, as Lucas *et al.* (1993) point out, plasmodesmata have evolved so as to quickly down-regulate traffick in case of wounding. Without such regulatory capabilities the symplast would quickly die. Callose plugging of plasmodesmata in wounded cells is a phenomenon commonly observed by those who studied wound reactions in plants. We observed such plugging specifically after abrasion for purposes of mechanical virus inoculation of plants. Currier and Webster (1964) showed plasmodesmatal plugging (SEL decrease) can occur quickly (seconds) whereas reversal of the process may take days. The concept of down-regulation of plasmodesmatal SEL after wounding certainly contradicts the idea of virus ingress and infection spread from wounds caused by abrasion. The wound-ingress concept of virus infection should, therefore, be revisited. It seems highly unlikely that, if replication and/or cotranslational uncoating occurs in the wounded cell, the products of these processes can be exported to the rest of the symplast through effectively closed plasmodesmata. These are issues open to experimentation.

This discussion of plant virus infection will be limited to the infection process of single-stranded, plus sense RNA viruses (ss(+)RNA), the largest and best studied group of plant viruses. Although cell wall–virus interactions of Caulimo, Gemini, Phytoreo and Rhabdo viruses are similar to those observed for the ss(+)RNA viruses, they will not be considered here. For the sake of clarity the infection process is divided into three host–virus interactive stages as depicted in Fig. 1. Each stage is defined by highly specific interactions not encountered in the other two stages. This presentation will consider each of the three interactions and compare and assess the data supporting the contrasting points of view on the plant virus infection process. My hope is that where there are conflicting views, unifying concepts may be found through research.

III. LEVEL I INTERACTIONS: VIRUS DESTABILIZATION EVENTS AT THE CELL SURFACE

The first level interaction involves the first association of host and virus at the cell surface and terminates when the infectious entity enters the cell

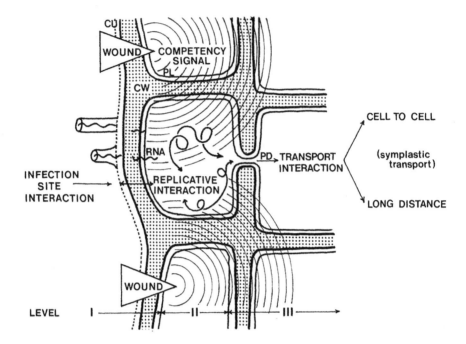

Fig. 1. The proposed processes involved in plant (+) sense RNA virus infection. Wounds made during inoculation generate signals that induce replication competency in the tissues (trigger high levels of host RNA-dependent-RNA-polymerases). Three interaction levels are shown. Level I interactions: A cascade of virus destabilizing events, initiated at the cell wall–virus interface (CU = cuticle, CW = cell wall) and terminated when the genomic RNA passes the plasmalemma (PL), comprises uncoating in the infection cycle. Similar destabilization is not repeated during subsequent invasion of the host. Note virions attached to the cuticle and the passage of RNA through the cell wall. Level II interactions: Translation and then replication is initiated utilizing a host membrane system (replication complex) as the backbone for genome transcription. Subsequent coat protein production and particle assembly removes most progeny RNA from the infection cycle since the necessary cell surface environment for uncoating progeny virions is absent in the symplast. Level III interactions: Cell-to-cell and long distance infection spread (invasion) are two distinct processes, mediated by transport of the replication complex which, through membrane fusion, initiate replication in compartments of the symplast hitherto untouched by the infection (PD = plasmodesma).

cytoplasm. The need for wounding of host tissues to establish successful plant virus infection is the most fundamental difference between plant viruses and their close relatives in the animal world.

The entrance of infectious virus into target cells through wounds is a generally accepted course of events in plant virus infections (Zaitlin and Hull, 1987). The wounds are thought to be portals for virus ingress into cells, or to expose specific attachment sites that differ from non-specific attachment sites found on unabraded cell walls. Although the virus literature is replete with examples providing unequivocal proof that wounding shortly before, during, or shortly after inoculation is necessary for successful plant virus infection, intensive study has not provided us with unequivocal proof for the conventional belief that viruses entering cells through wounds are the causal agent of infection (Matthews, 1981). Although wounds have been thought to function as infection sites, the possibility exists that wounding only induces competency to sustain virus replication in wound adjoining tissues. Induction of tissue competency includes, but may not be limited to, the production of increased amounts of a host RNA-dependent-RNA polymerase to be used in combination with the virally produced polymerase subunits to form virus specific polymerases in infected cells. Specific attachment sites, as in the case of bacterial viruses, or specific virus receptors on cell membranes, as in the case of animal viruses, have not been identified as yet for plant viruses. However, attachment of virus both to cell wall cuticulae and wound sites after mechanical inoculation of plants has been observed (Herridge and Schlegel, 1962; Gerola *et al.*, 1969; Gaard and de Zoeten, 1979). This early interaction between a plant host and a virus has been shown to be non-specific, non-saturable, and for the most part non-productive (Gaard and de Zoeten, 1979). The non-productivity of this attachment is exemplified by the extremely low specific infectivities [no. of lesions/particle $(3 \times 10^{-2}$ to $10^{-6})$] observed for plant viruses in comparison with bacterial and animal viruses. The issue of receptors in plant virus infection should be revisited in light of newer technologies now available for such studies (Bass and Greenberg, 1992). Statistical analysis of dilution experiments showed that theoretically one virus particle (one complete genome) can give rise to a single infection (Lauffer and Price, 1945). It must, therefore, be the inefficiency of the inoculation methods used for plant viruses, in part due to the lack of specific receptors, and the genome inactivation by omnipresent RNases released by the inoculation induced abrasions, that are responsible for the low specific activities observed for these viruses.

After inoculation, cytoplasmic disassembly is a necessity if intact virus enters through wounds. The phenomenon of cotranslational disassembly as first described by Wilson and coworkers (Wilson, 1984a, b; Wilson and Shaw, 1985; Turner *et al.*, 1987) could account for cytoplasmic uncoating. Cotranslational disassembly is the orderly disassembly of a virion that occurs when the cellular machinery initiates protein synthesis on a partially exposed viral translation initiation sequence. Progressive translation of the RNA by the ribosome would lead to further disassembly. However, the model that we propose challenges the presumption that virus causing the infection enters through wounds and that uncoating is in its entirety within the cellular boundary.

The apoplastic cell wall environment at the site of these first interactions becomes increasingly rich in coat protein (Cp) after inoculation. There is also apoplastic transport of the released Cp or its degradation products, both to the vascular bundle and to the plasma membrane via the cell wall (Gaard and de Zoeten, 1979) and the ectodesmata (Merkens *et al.*, 1972) which provide for solute transport from outside the cell to the plasma membrane.

Destabilization of plant viruses as a prelude to eventual uncoating is initiated in the lipids of the cuticulae at the attachment sites (de Zoeten, 1976; Durham, 1978). The destabilization process of virus attached to plant cells proceeds at relatively high speeds and is completed within 10–15 min after inoculation (Gaard and de Zoeten, 1979). Wu *et al.* (1994) reported similar results when they used more sophisticated techniques (PCR) which allowed them to substantially reduce the concentrations of the tobacco mosaic virus (TMV) inoculum. They showed that within 3 min of protoplast inoculation, uncoating had proceeded from the 5′ terminus of the viral RNA to a position approximately 4635 nucleotides away from the 5′ end. Interestingly, the uncoating of the message for the movement protein occurred much later, approximately 45 min into the experiment.

Ralton *et al.* (1986) addressed the problem of how the principles of signal recognition in animals can be applied to plants. Their conclusion was that although cyclic AMP, protein kinase, Ca^{2+} channels, calmodulin, and coated pits have all been described in plants, their roles in signal transduction in plants is essentially unknown. In a few plant host–pathogen systems the signal molecules (elicitors) are either of host or pathogen cell wall origin, released through enzymic action (Darvill and Albersheim, 1984). It is currently unknown whether a viral signal emanates from the attachment site, or possibly from the virus that entered the wounded cell. It is also unknown what would constitute such a signal and whether a signal is necessary at that stage of the infection process.

The observation that protoplasts can be infected by plant viruses apparently contradicts the necessity of a cell wall or its components for the initiation of uncoating. However, Burgess *et al.* (1973) have shown unequivocally that protoplasts are mortally wounded cells that begin to regenerate cell wall material within five minutes after enzyme removal. It is therefore erroneous to believe that protoplasts do not have cell wall material abutting their plasma membranes when inoculated. Moreover, if protoplasts are indeed mortally wounded cells, the results of infection studies with protoplasts are likely to be artifactual in comparison with the natural infection process.

A similar contradiction of the premise that uncoating is initiated at the cell surface was published by Turner *et al.* (1987) showing that injection of *Xenopus* oocytes with TMV particles resulted in the production of TMV non-structural proteins. This was interpreted to indicate that uncoating of TMV occurred within the cytoplasm of *Xenopus* oocytes. However, Hecht and Schlegel (personal communication) did similar experiments in 1966 with isolated tobacco leaf hair cells that were microinjected with TMV. It was not possible to prevent TMV from adhering to the outside of the glass needles used to microinject.

Moreover, upon retraction of the needle the added pressure of the micro-injected material resulted in small droplets of cellular contents protruding from the wound site of the injected cells (onto their surface). Therefore, Hecht and Schlegel could not conclude unequivocally that internal uncoating took place. Since no mention is made by Turner *et al.* (1987) of precautions taken to prevent such occurrences with *Xenopus* oocytes, we assume that they were not taken and that the infecting TMV could have been uncoated on the cell surface. A report on the cytoplasmic cotranslational uncoating of true and pseudo-TMV particles (Plaskitt *et al.*, 1988) in abraded epidermal cells provides neither the statistical analysis nor the cytological detail to adequately support the conclusions of the authors.

The next important question seems to be whether a wound site for the initiation of infection provides a competitive advantage over virus-cell wall interactions, functioning to initiate the infection. Depending on the severity of the wounding, decompartmentalization of cellular enzymes, including the release of RNases, proteases, and phenoloxidases, are taking place in the wound site. Increases of phenolics around and in wound sites are well documented reactions of plants to wounding (Ralton *et al.*, 1986). The importance of phenolics and their oxidation products for virus inactivation was shown by Mink (1965). Virions in wounds are therefore directly exposed to the dangers of these plant defense triggered compounds, whereas virus attached to cell walls at some distance from the abrasive created wounds may, because of their placement, escape these dangers. Wounding also collapses, through loss of turgor or by induction of callose deposition in the plasmodesmata (Lucas *et al.*, 1993), particulate communication channels (plasmodesmata) between wounded and surrounding cells. The wounded cell seems therefore not the best environment for the initiation of infection.

Over the last few years amino acid sequence similarities have been uncovered in the non-structural proteins (helicases, polymerases, etc.) of numerous viruses crossing the boundaries of the animal and plant kingdoms (Goldbach, 1986). Such similarities, whether the result of convergent or divergent evolution, argue for the existence of functional similarities. Although these sequence similarities are not found for the structural proteins, with few exceptions (satellite tobacco necrosis virus), all capsid proteins are composed of 180 eight-stranded antiparallel b–Barrels (ESABs) arranged with $T = 3$ symmetry or pseudo $T = 3$ symmetry. Thus, preservation of capsid structure seems a reality. It is therefore meaningful to examine the early virus–host interactions of the ss($+$)RNA animal viruses and compare them with what we know about early stages of plant virus–host interactions.

In vertebrate (Crowell and Lonberg-Holm, 1986), invertebrate, phyco (Meints *et al.*, 1984) and bacterial (Lewin, 1977) virology, where uncoating has been studied most, the process initiates at the cell surface and is completed there or close to the surface. During the infection by many picorna viruses, an eclipse period is observed. On the basis of virus inactivation studies, the physical location of the infectious entity at the time of eclipse must be on or close to the cell surface (Holland, 1962; Chan and Black, 1970). Mandel (1967) in his studies of this virus concluded that uncoating must be initiated at the cell surface, but taken to completion intracellularly by means of pinocytotic movement. Mandel, however, left the possibility open that mechanisms of infection other than viropexis (pinocytotic movement of virus) were operable as well (Dales, 1965; Cocking, 1970). De Sena and Mandel (1971) presented evidence that destabilization of the polio virus particle as a prelude to genome release occurred at the cell surface. Direct penetration of cells by polio virus has been observed by others (Dunnebacke et al, 1969), but has not been confirmed.

In 1986, Olsnes and others (Olsnes et al., 1986; Crowell and Lonberg-Holm, 1986) summarized our knowledge of animal virus attachment and entry into cells. The model for picorna virus infection at the time could be summarized as follows. The virus attaches to specific receptors in the cell membrane and the complexes are taken into endosomes (endocytic vesicles) and are then acidified to a pH at or below 6. The virion is altered by the acidic pH to an amphipathic intermediate structure, inserts itself into the vesicle membrane and then releases the genomic RNA into the cytoplasm. Even if this model turns out not to be universal, some parts of it (i.e. membrane passage of genomic RNA) will surely figure in most plant viral infection processes.

Since that time, important advances have been made in our understanding of picorna virus uncoating. Several viral receptors located on the cell membranes have been cloned and have been shown to be members of the immunoglobulin supergene family in animals. Whatever the sensitivity of different picorna viruses to low pH may be, low pH is "a parameter that is physiologically relevant to the entry process" (Hoover-Litty) and Greve, 1993) of these viruses. These results suggest that viral destablization during uncoating occurs in distinct stages and the authors speculate that incremental destabilization of the virion by the receptor acting in concert with other factors would be sufficient to effect uncoating in vivo. If immunoproteins form the building blocks of animal virus receptors it is not surprising that receptors have not been identified to play a role in plant virus infection since most probably plants, lacking a functional immune system, lack the ability to produce these proteins. Therefore, even if this picornavirus model is not universal, some parts of it (e.g. membrane passage of genomic RNA) will surely figure in most viral infection processes.

In plants, the cell wall pH is generally at or below 6, a pH needed for some picorna viruses to change their hydrophilic character to an amphipathic conformation so that pH-dependent integration of the particle coat protein with endosomal membranes can occur (Olsnes et al., 1986). Therefore, plants may not need the endosome (pinocytotic vesicles) to accomplish the conversion of the virus coat from hydrophilic to amphipathic. The ability of plant cells to form pinocytotic vesicles is still controversial. Thus, if low pH is needed for successful virus infection (in analogy with picornaviruses) the required pH is present in the various environs of the initial site of plant–virus interactions, obviating the apparent need for endosome formation and for intracellular uncoating. On the other hand, if pinocytosis in the form of endosome formation is needed for successful infection, currently there is no compelling evidence suggesting that such a mechanism is foreign to plant cells. Moreover, the ability of coat protein to destabilize membrane structure has been described for TMV-Cp (Banerjee *et al.*, 1981) and for the Cp of a protein-activated virus, alfalfa mosaic virus (AlMV) (Durham, 1978). The combination of cuticular destabilization of the particle and coat protein destabilization of the membrane may be the only requirements for membrane passage of the viral genome into plants. This is a distinct possibility since plant virus inocula have higher infectivities, when made in neutral to slightly alkaline buffers than when made in acid buffers. Some picorna viruses, however, do not have the requirement for a low pH to infect successfully and are inhibited by the same low pH that enhances polio infection.

The low specific infectivity of plant viruses noted earlier and the lack of discriminatory ability of cell walls in regard to attachment of pathogens versus non-pathogens in other plant host–parasite systems (Sequeira *et al.*, 1977; Watley and Sequeira, 1981; Sequeira, 1984) suggests, in our opinion, that the much more subtle discrimination between infecting and non-infecting virus particles (the latter constituting the bulk of the inoculum) cannot be accomplished by the cell wall. Regardless of the inoculum concentration, the initial interaction (destabilization) of all virions in an inoculum with the cell surface will be similar. The infecting particle will, therefore, have been subjected to most, or at least some of the conversions that the bulk of the inoculum underwent. The infecting particle (RNA), however, must in this "race" to the plasma membrane, remain genetically intact and escape the degradation that is the fate of the bulk of the inoculum. How that is accomplished other than maybe by default is currently unknown.

I have suggested that plant virus infection is initiated by destabilization of the virions at the first virus–host interface. The destabilization begins a cascade of events, which has both non-specific (early) and specific (late) components. The process encompasses an amalgamation of the destabilizing forces and interactions described by Durham (1978) and culminates with the passage of the viral genome into the host symplast. Although several mechanisms could be postulated to account for such genome passage, the actual mechanism is currently unknown. One possibility is that several Cp units produced during destabilization may reconstitute themselves during Cp-membrane interaction into a form, e.g. a lock washer disc conformation in the case of TMV, that allows RNA to cross the membrane. Thus a receptor may be created which is both specific for the RNA at hand and saturable. The conversion of the specific attachment site into an infection site and then into an infective center would result in a productive infection.

Whether or not the viral genome is devoid of Cp at membrane passage or associated with remnants of attached Cp seems to be immaterial for subsequent translational activity. It has been shown (Wilson, 1984a, b; Wilson and Watkins, 1984) that exposed initiation sequences on viral RNA will be utilized and that normal ribosome translocation along the RNA will strip the particle from its protein. This phenomenon was triggered *in vitro* by exposing particles to alkaline buffers although such environments do not exist *in vivo*. Nevertheless, cotranslational disassembly could account for the final stripping of partly uncoated virus entering the cell at the end of the cascade of destabilizing events triggered by the virus–cell wall interactions. This possibility might unify the two seemingly apposing views of uncoating. As it stands now, the *in vivo* work (Plaskitt *et al.*, 1988) claiming the isolation and visualization of striposome-like complexes is difficult to interpret in view of the equivocal electron microscope results and the lack of controls using serologically specific electron microscopy (SSEM) with anti-ribosomal sera to visualize the alleged striposome.

In this regard, the apparent need for coat protein-free RNA to initiate replication and the inhibitory effect of this protein on the process is significant. Horikoshi *et al.* (1987) showed unequivocally that brome mosaic virus replication is inhibited by its own coat protein. Spatial and or temporal separation of the replication function from coat protein production and virus assembly may, therefore, be a necessity. Thus, scavenging the last remnants of Cp from the viral RNA may be accomplished by the cotranslational uncoating process.

IV.   LEVEL II INTERACTIONS: THE FIRST CYCLE OF REPLICATION

Level II interactions (Fig. 1) encompass all activities involving the first cycle of replication from the release of the virion RNA into the cytoplasm until systemic invasion commences. During the level II interactions, virus-coded polymerase subunits are thought to combine with host RNA-dependent-RNA polymerase triggered by the wounding of the host during inoculation (induction of cell competency) to form a functional virus-specific RNA-dependent-RNA polymerase (Candresse *et al.*, 1986) (replicase). Repeated cycles of replication then produce progeny virus. It is unclear whether plant viruses regulate the ratio of replicatively used template to that used translationally, as do certain bacterial and animal viruses (Zaitlin and Hull, 1987). However, the temporal and spatial relationship of viral replication and translation products in infected cells is becoming increasingly clear (Ishikawa *et al.*, 1991; Reinke and de Zoeten, 1991).

It has been assumed for some time that plant viral polymerases capable of transcribing plant viral RNAs were intrinsic proteins of some host membrane (Brishammer, 1970; Bradley and Zaitlin, 1971; Zaitlin *et al.*, 1973; Brishammer and Juntti, 1974; Zabel *et al.*, 1974; Fraenkel-Conrat, 1976). The allotypic behaviour of most isolated viral polymerases when released from membranes is a well-known phenomenon of intrinsic membrane enzymes and could account for the loss of template specificity of viral polymerases when isolated (Brishammer, 1970; Bradley and Zaitlin, 1971; Zaitlin *et al.*, 1973; Brishammer and Juntti, 1974; Zabel *et al.*, 1974; Fraenkel-Conrat, 1976).

The occurrence of membrane-associated dsRNA has been established for TMV (Brishammer, 1970; Zaitlin *et al.*, 1973; Brishammer and Juntti, 1974), cowpea mosaic virus (CpMV) (Assink *et al.*, 1973; de Zoeten *et al.*, 1974) and turnip yellow mosaic virus (TYMV) (La Fleche *et al.*, 1972). We showed that both pea enation mosaic virus (PEMV)-specific dsRNA and viral polymerase are associated with a host-generated membrane system originating in the nucleus (de Zoeten *et al*, 1972, 1976). Since none of the replication products themselves are mobile, transport of the infectious entity must be achieved in a passive manner following the modes of transport present in the symplast.

It has been shown for several plant viruses (Lockhart and Semancik, 1969; Dawson and Schlegel, 1976; Dawson, 1978) that actinomycin D administered shortly before, during or immediately after inoculation inhibits an early step in virus replication. This may indicate that viral infection depends on an unstable constitutively expressed host protein; that communication between the viral genome and the host genome is a prerequisite for successful infection; or that the host's wound response resulting in cell competency is inhibited. This early interaction results somehow in the formation by the host of a membrane-associated replicase system used in viral replication.

V. LEVEL III INTERACTIONS: CELL-TO-CELL AND LONG-DISTANCE SPREAD OF THE INFECTION

Level III interactions deal mainly with the invasion of the symplast by the infectious entity. For an exhaustive and elegant assessment of the spatial and temporal relationship of the transport function to virus invasion, the reader is referred to reviews by Atabekov and Dorokhov (1984) and Atabekov and Talianski (1990).

If competency to replicate is produced by wounding, it seems clear that competency to transport is induced by virus replication itself and that there are proteins produced that "open the gates" (Atabekov and Dorokhov, 1984), that is, the plasmodesmata for transport of the infectious agent. In the past few years a veritable watershed of research reports addressing viral movement proteins, their functional capabilities, their effect when expressed as transgenes, their ability to complement movement protein mutants and their universality in providing transport functions to unrelated viruses in transit and as transgenes has been published (reviewed by Atabekov and Talianski, 1990; Deom *et al.*, 1992). Where localization and mutational studies have been initiated, the wild type TMV-30K movement protein was found associated with plasmodesmata while the mutated movement protein (MP Δ 3–5) associated itself to a much lesser extent with plasmodesmata and did not change the SEL to the same extent as the wild type (Lapidot *et al.*, 1993). Similar studies with the putative movement proteins from a Geminivirus showed increased plasmodesmatal SEL for cells treated with these proteins indicating the universality of the effect of plant viral movement proteins on plasmodesmatal trafficking (Noueiry *et al.*, 1994). The current, and hopefully unifying, concept of the functionality of viral movement proteins holds that there are several interactive domains on these proteins that provide, among others, for ss-RNA binding, for alteration of SEL of plasmodesmata and possibly for host specificity. The transport of the infectious agent itself may passively follow the routes provided by the symplastic channels. However, virus actively participates in opening those routes (e.g. 30 K protein; Leonard and Zaitlin, 1982). Direct evidence for the transport role of the 30 K protein was provided first by Beachy's group (Zaitlin and Hull, 1987) which showed that strain LS-1 defective in this protein at restrictive temperatures will move in plants that are transgenic for the 30 K protein.

There are numerous reports on the presence of virus particles in plasmodesmata (Esau, 1967; de Zoeten and Gaard, 1969; Kitajima and Lauritis, 1969; Kim and Fulton, 1973). These observations by no means indicate the only form in which the viral infection spreads but rather that plasmodesmata can accommodate the transport of virus particles.

When time course studies of movement of the infectious agent were performed it became clear that the cell-to-cell transport is a relative early function, taking place before rather than after mature virions have been formed. The infection of epidermal cells generally moves to the underlying cells of inoculated leaves several hours before the first particles can be found (Matthews, 1981). It is perhaps noteworthy in this connection to realize that the TMV 30 K protein implicated in transport of the infection is ephemeral and is only synthesized early in infection after which synthesis declines (Watanabe *et al.*, 1984). Furthermore, Nishiguchi *et al.* (1978, 1980) showed clearly that the presence of virions in primary infected cells is not, in itself, sufficient for cell-to-cell spread of the infection. Also, the infection of a temperature-sensitive mutant (35-IV; Dawson, 1983) defective in particle formation shows good invasive properties in our hands, notwithstanding the lack of virion production. There is currently a large body of evidence suggesting that virus particle formation is not a *conditio sine qua non* for cell-to-cell spread of the infection, whereas there is no evidence that long-distance transport is dependent on the presence of virions. The new Umbravirus group of viruses lack coat protein, but can move efficiently over both short and long distance when inoculated onto hosts without their usual Luteovirus counterparts (helpers)

(Demler *et al.*, 1995). No generalizations can be made at this time as to whether it is the coat protein alone, the viral coat protein gene alone, or combinations of these that are necessary for long-distance transport. Furthermore, host genotype and environmental factors may also play a role in long-distance transport (Xiong *et al.*, 1993).

The principal difference between short-distance (cell-to-cell) and long-distance transport of the infectious agent has been noted in studies by Atabekov and coworkers who used complementation of transport function mutants *in vivo* (reviewed in Atabekov and Dorokhov, 1984; Atabekov and Talianski, 1990). Their conclusions were confirmed by Takamatsu *et al.* (1987) who replaced the majority of the TMV coat protein gene with the CAT (chloramphenicol acetyltransferase) gene and showed "that the coat protein and its coding sequence play at least no positive role in viral replication or in viral cell-to-cell movement" as Atabekov and Dorokhov (1984) found. Takamatsu *et al.* (1987) suggested that TMV coat protein plays an important role in long-distance transport since they did not observe spread of the CAT mutant or the coat protein deletion mutant to upper leaves of inoculated plants.

Dawson's group (personal communication) tested various deletions in the TMV *Cp* gene, within the context of an otherwise wild-type genome. The infections spread slowly and erratically from the inoculated leaves, with some secondary leaves escaping infection. Ahlquist's group (University of Wisconsin, Madison; personal communication) has tested several brome mosaic virus (BMV) *Cp* mutants, ranging from frame shifts involving a few bases to partial or complete deletions of the gene. Although the mutants replicated well in protoplasts, no systemic spread was seen in whole plants with any mutant lacking a functional *Cp* gene. We conclude from these experiments that long-distance transport of the infectious entity in these cases is dependent on the presence of viable coat protein provided either by the viral genome itself or by complementation. The natural complementation of the impaired long-distance transport function of the tobacco rattle virus (TRV) "NM" variants (the naked RNA of the long particle (Cadman, 1962; Lister, 1968)), by the short particle which carries the coat protein gene, should in retrospect have provided us with clues to the different requirements of short-and long-distance infection spread in regard to viable coat protein.

The lack of messenger activity of viroid RNA and its relatively good invasive powers (Dickson, 1979) remains an enigma when we consider the requirement for coat protein of long-distance transport of virus infection.

In summary, arguments have been presented for the view that plant virus infection begins at the cell surface where uncoating is initiated. Uncoating is completed shortly thereafter when viral RNA emerges from the plasmalemma in the cell proper to initiate translation and replication. This, I consider, is a single event in plant virus infection, not repeated during the invasion of the symplast, since progeny virus is apparently non-essential in the spread of infection. Although the results of Saito *et al.* (1990) are generally taken to indicate that intact virus plays a role in long-distance transport, nothing in their work indicates this unequivocally. Saito and coworkers showed, as have many others, that intact, assembly-competent coat protein is needed for long-distance transport and furthermore, that an intact assembly-competent, assembly origin is necessary for successful long-distance transport. Their conclusion (Saito *et al.*, 1990 p. 334) is tentative as to the involvement of virus particles: "Thus, virus particles may play a pivotal role in the movement, although the possibility cannot be completely eliminated that other complexes such as informasome-like ribonucleoprotein complexes (Dorokhov *et al.*, 1984), which might require the coat protein and assembly origin for their formation are also involved." In this viewpoint the latter part of their conclusion is favored over the former.

Returning now to the main discussion, the environment of the cytosol does not provide the initial destabilizing forces necessary for virion uncoating as found in the cuticle. Furthermore, the regulatory mechanisms to discriminate in the cytosol between virions that need to be uncoated for successful systemic invasion to take place and those that should remain intact as a necessity for virus survival have never been described, studied, or proven to exist! It is argued here that fully assembled virions, once formed, are lost to infection spread within the host, similar to many resting spores in other host–pathogen systems.

Complete cytosolar continuity of the symplast throughout the plant via plasmodesmata and structures such as sieve pore plates in phloem elements allows actual particulate transport between the compartmentalized parts of the healthy plant symplast. This continuity, possibly enhanced and maintained by virus coded proteins (movement proteins) that overcome plant defenses, allows cell-to-cell communication in virus-infected plants to a degree far exceeding the communication abilities present in animal tissues. As opposed to animal virus invasion, which depends on repetitive cycles of virus particle release (lysis) and cell infection, plant viruses seem to have developed invasion strategies adapted to the continuum presented by the plant symplast. Systemic virus invasion seems to depend on the spread of the infectious entity in a form different from the intact virion.

The first experiments indicating that mature virions do not represent the form in which the infection is spread were performed by Dawson *et al.* (1975). In these experiments the lower leaves of plants mechanically inoculated with TMV were kept at 25°C while the upper leaves were maintained at 5°C. Under such a regime of differential temperature treatment (DTT) it was determined that virus replication occurred in the leaves kept at 25°C and not in those at 5°C. When the tops were detached and transferred to 25°C virus replication ensued at near synchrony, notwithstanding the fact that at the time of severance no infectivity could be recovered from the tops of these plants.

The conclusion that the TMV genome reached the top of the plant in a form different from that allowing recovery of infectivity (presumably the virions produced in the lower part of the plant) is obvious. The form such an entity would take, however, is less obvious. In 1984 Atabekov and Dorokhov reviewed the evidence for the involvement of "virus-specific informosome-like ribonucleoprotein (vRNP)" in transport. The reader is referred to their review for details of their concept of transport of infection in plants. They provided evidence that the form in which the infectious entity travels is a cytoplasmic ribonucleoprotein (RNP) particle built of cellular components in complex with the viral RNA. One form of such vRNP is associated with membranes, probably the membranes of the replication complex used as the backbone for ss(+)RNA replication as described by de Zoeten *et al.* (1972, 1976). To our knowledge, they have not provided EM images of vRNPs *in situ*. In the case of TMV, such images are difficult to produce, since the form of the replication complex of TMV *in situ* is currently uncertain. Replication complexes of other viruses that are morphologically distinct from normal cell constituents have been described. For pea enation mosaic virus, de Zoeten and Gaard (1983) showed that the replication complex (vesicular material) could be transported in the phloem from cells that produce this structure, the companion and adjoining parenchyma cells, to cells that do not produce it, the mature phloem elements which lack nuclei. Localization of replication within or on membrane systems would provide the protection of the nucleic acids ((+)RNA, (−) RNA, dsRNA) needed for long-distance transport through phloem elements where enzyme compartmentalization may have broken down during ontogeny.

I proposed here that systemic invasion of the symplast is mediated by the transport of these self-contained replication complexes containing vRNPs. Replication is initiated wherever the environment is conducive to membrane fusion resulting in the reverse process that originally formed the replication complex. Esau and Hoefert (1972) inferred from their studies of beet western yellows virus that such membrane fusion may occur between membranes in the cytoplasm (the replication complex) and the nuclear membrane. Unequivocal proof for the infectious nature of these complexes will, however, be hard to obtain.

VI. CONCLUSION

The summarized evidence purports the following.

1. Virus uncoating is a single event in virus infection that originates at the first plant−virus interface, the cell wall, arguing that at this interface a cascade of events is initiated that eventually provides the first translatable viral message to the cell.

2. Virus−host communication at this initial stage of infection results in the formation of a virus replication complex of host (cell competency) and viral origin.

3. Based on coat protein requirements, short-and long-distance transport

of the infection are two distinct processes that do not require the transport of actual virions.

4. vRNPs or replication complexes are the form in which ss(+)RNA virus infection spreads throughout the plant symplast.

In presenting "Another Point of View" I have attempted to reconcile the current knowledge of plant virus infection in general terms with the enduring knowledge of the virus infection process developed earlier or in the other branches of virology. I have emphasized the principal differences between the plant symplast and the animal cell and propose on the basis of this difference a scenario for systemic virus invasion of plants that in my view provides a logical alternative to the currently held views.

At this time we have only circumstantial or inferential evidence for a number of the phenomena referred to in this review. However, they are now open to research approaches not available in years past. Among these are: (1) new biochemical, immunological and molecular (rDNA) approaches to answer questions, in regard to plant virus receptors; (2) directed mutagenesis of cloned viral cDNAs makes it possible to localize gene functions on the viral genomes; and (3) the ability to manufacture transgenic plants that can express any genomic sequence imaginable enabling gene complementation studies in trans. The combination of these tools in our studies of virus infection will allow us to answer many of the vexing questions of the past.

ACKNOWLEDGEMENTS

The author wishes to express his appreciation to the many colleagues that read the manuscript over the years and that made many helpful suggestions.

REFERENCES

Assink, A. M., Swaans, H. and Van Kanitmen, A. (1973). The localization of virus-specific double stranded RNA of cowpea mosaic virus in sub-cellular fractions of *Vigna* leaves. *Virology* **58**, 384–391.

Atabekov, J. G. and Dorokhov, Y. L. (1984). Plant virus-specific transport function and resistance of plants to viruses. *Advances in Virus Research* **29**, 313–364.

Atabekov, J.G. and Talianski, M. E. (1990). Expression of a plant virus-coded transport function by different viral genomes. *Advances in Virus Research* **38**, 201–249.

Banerjee, S., VanDenbrandeu, M. and Ruysschaert, F. M. (1981). Interaction of tobacco mosaic virus protein with lipid membrane systems. *FEBS Letters* **133**, 221–229.

Bass, D. M. and Greenberg, B. (1992). Strategies for the identification of icosahedral virus receptors. *Journal of Clinical Investigation* **89**, 3–9.

Bradley, D. W. and Zaitlin, M. (1971). Replication of tobacco mosaic virus. II. The *in vitro* synthesis of high molecular weight virus-specific-RNA's. *Virology* **45**, 192–199.

Brishammer, S. (1970). Identification and characterization of an RNA replicase from TMV infected tobacco leaves. *Biochemical Biophysical Research Communications* 41, 506–511.

Brishammer, S. and Juntti, N. (1974). Partial purification and characterization of soluble TMV replicase. *Virology* 56, 245–253.

Burgess, J., Motoyoshi, F. and Flemming, E. N. (1973). Effect of poly-L-ornithine on isolated tobacco mesophyll protoplasts: evidence against stimulated pinocytosis. *Planta III*, 199–208.

Cadman, C. H. (1962). Evidence for association of tobacco rattle virus nucleic acid with a cell component. *Nature* 193, 49–52.

Candresse, T., Mouches, C. and Bové, J. M. (1986). Characterization of the virus encoded subunit of turnip yellow mosaic virus RNA replicase. *Virology* 152, 322–330.

Chan, V. F. and Black, F. L. (1970). Uncoating of polio virus by isolated plasma membranes. *Journal of Virology* 5, 309–312.

Cocking, E. C. (1970). Virus uptake, cell wall regeneration and virus multiplication in isolated plant protoplasts. *International Review of Cytology* 28, 29–124.

Crowell, R. and Lonberg-Holm, K. (eds) (1986) Introduction and overview. *In* "Virus Attachment and Entry into Cells", pp. 1–9. American Society for Microbiology, Washington, DC.

Currier, H. B. and Webster, D. H. (1964). Callose formation and subsequent disappearance: studies in ultrasound stimulation. *Plant Physiology* 39, 843–847.

Dales, S. (1965). Penetration of animal viruses into cells. *Medical Virology* 7, 1–43.

Darvill, A. G. and Albersheim, P. (1984). Phytoalexins and their elicitors — a defense against microbial infection in plants. *Annual Review of Plant Physiology* 35, 243–275.

Dawson, W. O. (1978). Time-course of cowpea chlorotic mottle virus RNA replication. *Intervirology* 9, 119–128.

Dawson, W. O. (1983). Tobacco mosaic virus protein synthesis is correlated with double-stranded RNA synthesis and not single-stranded RNA synthesis. *Virology* 125, 314–323.

Dawson, W. O. and Schlegel, D. E. (1976). The sequence of inhibition of tobacco mosaic virus synthesis by actinomycin D, 2-thiouracil, and cycloheximide in a synchronous infection. *Phytopathology* 66, 177–181.

Dawson, W. O., Schelegel, D. E. and Lung, M. C. Y. (1975). Synthesis of tobacco mosaic virus in intact tobacco leaves systemically inoculated by differential temperature treatment. *Virology* 65, 565–573.

Demler, S. A., de Zoeten, G. A., Adam, G. and Harris, K. F. (1995). Pea enation mosaic virus: Properties and aphid transmission. In "The Viruses: Plant Viruses". Plenum Press, New York.

Deom, C. M., Lapidot, M. and Beachy, R. N. (1992). Plant virus movement proteins. *Cell* 69, 221–224.

De Sena, J. and Mandel, B. (1971). Studies on the *in vitro* uncoating of polio virus. II. Characteristics of the membrane modified particle. *Virology* 28, 554–566.

de Zoeten, G. A. (1976). Cytology and physiology of penetration and establishment. *In* "Encyclopedia of Plant Physiology" (New Series, Vol. 4), (R. Heitefuss and P. H. Williams, eds) pp. 129–149. Springer-Verlag, Berlin, Heidelberg, New York.

de Zoeten, G. A. and Gaard, G. (1969). Possibilities for inter- and intra-cellular translocation of some icosahedral plant viruses. *Journal of Cell Biology* 40, 814–823.

de Zoeten, G. A. and Gaard, G. (1983). Mechanisms underlying systemic invasion of pea plants by pea enation mosaic virus. *Intervirology* 19, 85–94.

de Zoeten, G. A., Gaard, G. and Diez, F. B. (1972). Nuclear vesiculation associated with pea enation mosaic virus-infected plant tissue. *Virology* **48**, 638–647.

de Zoeten, G. A., Assink, A. M. and Van Kammen, A. (1974). Association of cowpea mosaic virus induced double stranded RNA with a cytopathological structure in infected cells. *Virology* **59**, 341–355.

de Zoeten, G. A., Powell, C. A. and Gaard, G. (1976). *In situ* localization of pea enation mosaic virus double stranded ribonucleic acid. *Virology* **70**, 459–469.

Dickson, E. (1979). Viroids: infections RNA in plants. *In* "Nucleic Acids in Plants" Vol. 2 (T. C. Hall and J. W. Davies, eds), pp. 183–193. CRC Press, Boca Raton, Florida.

Dunnebacke, T. H., Levinthal, J. D. and Williams, R. C. (1969) Entry and release of polio virus as observed by electron microscopy of cultured cells. *Journal of Virology* 4, 505–513.

Durham, A. C. H. (1978). The roles of small ions especially calcium in virus disassembly and transformation. *Biomedicine* **28**, 307–313.

Esau, K. (1967). Anatomy of plant virus infection. *Annual Review of Phytopathology* **5**, 45–76.

Esau, K. and Hoefert, L. L. (1972). Development of infection with beet western yellows virus in sugarbeet. *Virology* **48**, 724–738.

Fraenkel-Conrat, H. (1976). RNA polymerase from tobacco necrosis virus in infected and uninfected tobacco. Purification of the membrane-associated enzyme. *Virology* **72**, 23–32.

Gaard, G. and de Zoeten, G. A. (1979). Plant virus uncoating as a result of virus-cell wall interactions. *Virology* **96**, 21–31.

Gerola, F. M., Bassi, M., Favali, M. A. and Betto, E. (1969). An electron microscopy study of the penetration of tobacco mosaic virus into leaves following inoculation experiments. *Virology* **68**, 380–386.

Goldbach, R. W. (1986). Molecular evolution of plant viruses. *Annual Review of Phytopathology* **24**, 289–310.

Gunning, B. E. S. and Robards, A. W. (eds) (1976). "Intercellular Communication in Plants: Studies on Plasmodesmata". Springer-Verlag, Berlin, Heidelberg.

Herridge, E. A. and Schlegel, D. E. (1962). Autoradiographic studies of tobacco mosaic virus inoculations on host and non host species. *Virology* **18**, 517–523.

Holland, J. J. (1962). Irreversible eclipse of polio virus by Hela cells. *Virology* **16**, 163–176.

Hoover-Litty, H. and Greve, J. M. (1993). Formation of rhino virus-soluble ICAM-1 complexes and conformational changes in the viron. *Journal of Virology* **67**, 390–397.

Horikoshi, M., Nakayama, M., Yamaoka, N., Furusawa, I. and Shishyama, J. (1987). Brome mosaic virus coat protein inhibits viral RNA synthesis *in vitro*. *Virology* **158**, 15–19.

Ishikawa, M., Meshi, T., Ohno, T. and Okada, Y. (1991). Specific cessation of minus-strand RNA accumulation at an early stage in tobacco mosaic virus infection. *Journal of Virology* **65**, 861–868.

Kim, K. S. and Fulton, J. P. (1973). Plant virus-induced cell wall overgrowth and associated membrane elaboration. *Journal of Ultrastructure Research* **45**, 328–342.

Kitajima, E. W. and Lauritis, J. A. (1969). Plant virions in plasmodesmata. *Virology* **37** 681–685.

LaFleche, C., Bove, C., Dupont, C., Mouches, C., Astier, T., Garnier, M. and Bové, J. M. (1972). Site of viral RNA replication in cells of higher plants: TYMV-RNA on the chloroplast outer membrane. *In* RNA viruses/Ribosomes, Proceedings of the 8th Meeting of the Federation of European Biochemical Societies (H. Bloemendaal *et al.* organizers) pp. 43–71. North Holland, American Elsevier, New York.

Lapidot, M., Gofny, R., Ding, B., Wolf, S., Lucas, W. J. and Beachy, R. N. (1993). A dysfunctional movement protein of tobacco mosaic virus that partially modifies the plasmodesmata and limits virus spread in transgenic plants. *Plant Journal* 4, 959–970.

Lauffer, M. A. and Price, W. C. (1945). Infection by viruses. *Archives of Biochemistry* 8, 449–468.

Leonard, D. A. and Zaitlin, M. (1982). A temperature sensitive strain of tobacco mosaic virus defective in cell-to-cell movement, generates an altered viral-coded protein. *Virology* 117, 416–424.

Lewin, B. (1977). Gene Expression, Vol. 3, Plasmids and Phages, John Wiley, New York, Sidney, London, Toronto.

Lister, R. M. (1968). Functional relationships between virus specific products of tobacco rattle virus. *Journal of General Virology* 2, 43–58.

Lockhart, B. E. L. and Semancik, J. S. (1969) Differential effects of actinomycin-D on plant virus multiplication. *Virology* 39, 362–365.

Lucas, W. J., Ding, B. and Van Der Schoot, C. (1993). Tansley Review No. 58. Plasmodesmata and the supracellular nature of plants. *New Phytologist* 125, 435–576.

Mandel, B. (1967). The relationship between penetration and uncoating of polio virus in Hela cells. *Virology* 31, 701–712.

Matthews, R. E. F. (1981). "Plant Virology" Academic Press, New York.

Meints, R. H., Lee, K., Burbank, D. E. and Van Etten, J. L. (1984). Infection of a *Chlorella*-like alga with the virus, PBCV-1: Ultrastructural studies. *Virology* 138, 341–346.

Merkens, W. S. W., de Zoeten, G. A. and Gaard, G. (1972). Observations on ectodesmata and the virus infection process. *Journal of Ultrastructure Research* 41, 397–405.

Mink, G. I. (1965). Inactivation of tulare apple mosaic virus by D-quinones. *Virology* 26, 700–707.

Nishiguchi, M., Motoyoshi, F. and Oshima N. (1978). Behaviour of a temperature sensitive strain of tobacco mosaic virus in tomato leaves and protoplasts. *Journal of General Virology* 39, 53–61.

Nishiguchi, M., Motoyoshi, F. and Oshima N. (1980). Further investigation of a temperature sensitive strain of tobacco mosaic virus: its behavior in tomato leaf epidermis. *Journal of General Virology* 46, 496–500.

Noueiry, A. O., Lucas, W. J. and Gilbertson, R. L. (1994). Two proteins of a plant DNA virus coordinate nuclear and plasmodesmatal transport. *Cell* 76, 925–932.

Olsnes, S., Madshus, I. H. J. and Sandvig, K. (1986). Entry mechanisms of picornaviruses. *In* "Virus Attachment and Entry into Cells", pp. 171–181. American Society for Microbiology, Washington, DC.

Plaskitt, A. K., Watkins, P. A. C., Heat, D. E., Gallie, D. R., Shaw, J. G. and Wilson, T. M. A. (1988). Immunogold labeling locates the site of disassembly and transient gene expression of tobacco mosaic virus-like pseudo virus particles *in vivo*. *Molecular Plant Microbe Interactions* 1, 10–16.

Ralton, J. E., Howlett, B. J. and Clarke, A. E. (1986). Receptors in host–pathogen interactions. *In* "Hormones, Receptors and Cellular Interactions in Plants" (C. M. Chadwick and D. R. Garrod, eds) pp. 281–318. Cambridge University Press, Cambridge.

Reinke, J. K. and de Zoeten, G. A. (1991). *In situ* localization of plant viral gene products. *Phytopathology* **81**, 1306–1314.

Robards, A. W. and Lucas, W. J. (1990). Plasmodesmata. *Annual Review of Plant Physiology* **41**, 369–419.

Saito, T., Yamanaka, K. and Okada, Y. (1990). Long distance movement and viral assembly of tobacco mosaic virus mutants. *Virology* **176**, 329–336.

Sequeira, L. (1984). Plant-bacterial interactions. *In* "Encyclopedia of Plant Physiology" (new series) Vol. 17 (H. F. Linskens and J. Heslop-Harrison, eds.) pp. 187, 211. Springer-Verlag, Berlin, Heidelberg, New York, Tokyo.

Sequeira, L., Gaard, G. and de Zoeten, G. A. (1977). Interaction of bacteria and host cell walls: its relation to mechanisms of induced resistance. *Physiological Plant Pathology* **10**, 43–50.

Shaw, J. G., Plaskitt, K. A. and Wilson, T. M. A. (1986). Evidence that tobacco mosaic virus particles disassemble cotranslationally *in vitro*. *Virology* **148**, 326–336.

Takamatsu, N., Ishikawa, M., Meshi, T. and Okada, Y. (1987). Expression of bacterial chloramphenicol acetyltransferase gene in tobacco plants mediated by TMV-RNA. *EMBO Journal* 6, 307–311.

Turner, P. C., Watkins, P. A. C., Zaitlin, M. and Wilson, T. M. A. (1987). Tobacco mosaic virus particles uncoat and express their RNA in *Xenopus laevis* oocytes: implications for early interactions between plant cells and viruses. *Virology* **160**, 515–517.

Watanabe, Y., Emori, Y. Ooshika, I., Meshi, T., Ohno, T. and Okada, Y. (1984). Synthesis of TMV-specific RNAs and proteins at the early stage of infection in tobacco protoplasts: transient expression of the 30K protein and its mRNA. *Virology* **133**, 18–24.

Watley, M. H. and Sequeira, L. (1981). Bacterial attachment to plant cell walls. *Recent Advances in Phytochemistry* **15**, 213–240.

Wilson, T. M. A. (1984a). Cotranslational disassembly of tobacco mosaic virus *in vitro*. *Virology* **137**, 255–265.

Wilson, T. M. A. (1984b). Cotranslational disassembly increases the efficiency of expression of TMV RNA in wheat germ cell-free extracts. *Virology* **138**, 353–356.

Wilson, T. M. A. and Shaw, J. G. (1985). Does TMV uncoat cotranslationally *in vivo*? *Trends in Biochemical Science* 10, 57–60.

Wilson, T. M. A. and Watkins, P. A. C. (1984). Cotranslational disassembly of a cowpea strain (Cc) of TMV: evidence that viral RNA protein interactions at the assembly origin block ribosome translocation *in vitro*. *Virology* **145**, 346–349.

Wu, X., Xu, Z. and Shaw, J. G. (1994). Uncoating of tobacco mosaic virus RNA in protoplasts. *Virology* 200, 256–262.

Xiong, Z., Kim, K. H., Giesman-Cookmeyer, D. and Lommel, S. A. (1993). The roles of the red clover necrotic mosaic virus capsid and cell-to-cell movement proteins in systemic infection. *Virology* **192**, 27–32.

Zabel, P., Weegen-Swaans, H. and Van Kammen, A. (1974). *In vitro* replication of cowpea mosaic virus RNA. I. Isolation and properties of the membrane-bound replicase. *Journal of General Virology* **14**, 1049–1055.

Zabel, P., Jongen-Neven, I. and Van Kammen, A. (1976). *In vitro* replication of cowpea mosaic virus RNA. Solubilization of membrane-bound replicase and the partial purification of the solubilized enzyme. *Journal of General Virology* **17**, 679–685.

Zaitlin, M. and Hull, R. (1987) Plant virus–host interactions. *Annual Review of Plant Physiology* **38**, 291–315.

Zaitlin, M., Duda, C. T. and Petti, M. A. (1973). Replication of tobacco mosaic virus. V. Properties of the bound and solubilized replicase. *Virology* **53**, 300–311.

The Pathogens and Pests of Chestnuts

SANDRA L. ANAGNOSTAKIS

The Connecticut Agricultural Experiment Station, New Haven, Connecticut 06504-1106, USA

I. INTRODUCTION

The name "Chestnut" refers to seven species of deciduous, nut-bearing trees found native and introduced throughout the world (Table I). American chestnut trees (*Castanea dentata*) were once so common in the Eastern United States that everyone who could get to the woods in the fall could count on nuts for roasting, and for stuffing their Thanksgiving turkey (Fig. 1). The wood is highly resistant to rot, and was used extensively for poles, fencing, building materials, and musical instruments (Table II). The bark was an important source of tannin for the leather tanning industry. Because chestnuts have been important to people all over the world for a very long time,

Advances in Botanical Research Vol. 21
incorporating Advances in Plant Pathology
ISBN 0-12-005921-5

TABLE I

Species in the genus Castanea

SECTION Castanea (three nuts per bur)
Castanea dentata (Marshall) Borkhausen — American Chestnut
Castanea sativa Miller — European Chestnut
Castanea mollissima Blume — Chinese Chestnut
Castanea crenata Siebold and Zuccarini — Japanese Chestnut
Castanea seguinii Dode — Chinese Dwarf Chestnut

SECTION Balanocastanon (one nut per bur)
Castanea pumila (Linnaeus) Miller
  variety *pumila* — Allegheny Chinquapin
  variety *ozarkensis* — Ozark Chinquapin

SECTION Hypocastanon [one nut per bur]
Castanea Henryi (Skan) Rehder & Wilson — Chinese Timber Chinquapin

TABLE II

Connecticut industries using milled chestnut wood in 1912[a]

NAME OF INDUSTRY	Quantity		Cost ($)	
	Feet, b.m.[b]	% of total	Average per 1000 ft	Total
Musical instruments	3 559 000	49.1	21.58	76 815.50
Planing mill products	839 500	11.6	46.48	39 017.00
Sash, doors, blinds and general mill work	683 480	9.4	37.61	25 704.15
Ships and boats	546 645	7.6	23.54	12 866.71
Miscellaneous	440 000	6.1	22.68	9 980.00
Clocks	285 000	3.9	19.02	5 420.00
Fixtures	245 500	3.4	23.20	5 696.50
Pro and scientific instruments	161 000	2.2	18.07	2 910.00
Boxes and crates	142 500	2.0	14.82	2 111.50
Wooden ware	135 000	1.9	13.56	1 830.00
Furniture	78 000	1.1	22.27	1 737.00
Machinery and apparatus, not electrical	44 975	0.6	23.84	1 072.30
Patterns	20 000	0.3	22.00	440.00
Laundry appliances	17 500	0.2	22.29	390.00
Agricultural implements	15 000	0.2	20.00	300.00
Vehicles and vehicle parts	12 800	0.2	25.00	320.00
Handles	10 000	0.1	18.00	180.00
Printing materials	5 800	0.1	35.00	203.00
Electrical machinery and apparatus	3 000	[c]	20.00	60.00
Totals	7 244 700	100.0	25.82	187 053.66

[a] Albert H. Pierson, Connecticut Agricultural Experiment Station Bulletin #174, 96 pp., New Haven, Connecticut.
[b] b.m. is "board measure" and refers to a board 1 inch × 12 inches × 12 inches.
[c] Less than 0.1 of 1%.

Fig. 1.   An American chestnut tree (*Castanea dentata* (Marsh.) Borkh.) in Scotland, Connecticut in 1905. This tree was described as a seedling, 103 years old, 27 inches in diameter, 83 feet tall, and occupying 900 square feet of crown space. Photographed by W. O. Filley, The Connecticut Agricultural Experiment Station, New Haven, Connecticut.

their problems have been the focus of many research reports and popular articles.

II.   INK DISEASE

In 1838 a root disease was reported on European chestnut in Portugal, and soon after this "ink disease" was found in several other parts of southern

Europe (Crandall *et al.*, 1945). The pathogens, *Phytophthora cinnamomi* Rands and *P. cambivora* (Petri) Buism. cause brownish-black lesions on the roots, and these lesions exude an inky-blue stain. Trees are killed when the stem is girdled, or when most of the roots are killed. Breeding with resistant, Asian chestnut trees has produced new resistant cultivars and rootstocks, and these, along with careful sanitation and site selection, allow chestnuts to still be grown in infected areas in Europe (Abreu, 1992; Breisch, 1992).

Ink disease is assumed to have entered the United States in the mid-1800s, and probably accounted for the recession of American chestnut from large areas in the Gulf and Mid-Atlantic states in the United States, and inland to the foothills and mountains of Mississippi, Alabama, Georgia, Tennessee, Maryland, Virginia, and the Carolinas (Crandall *et al.*, 1945). This chestnut destruction was occurring at a time when extensive clear-cutting was being practised, and the failure of the chestnuts to sprout in some places was noted but not pursued. *P. cinnamomi* and *P. cambivora* are both present in the US, but Crandall *et al.* (1945) concluded that only *P. cinnamomi* was causing ink disease of American chestnut and chinquapins.

The second disaster for American chestnuts was also an "imported" fungus disease. It was discovered in New York City in 1904, and within 50 years it had changed the appearance of our eastern forests.

III. CHESTNUT BLIGHT

Chestnut Blight, or Chestnut Bark Disease is caused by the fungus *Cryphonectria parasitica* (Murr.) Barr, formerly called *Endothia parasitica* (Murr.) And. & And. The fungus enters wounds, grows in and under the bark, and eventually kills the cambium all the way around the twig, branch, or trunk (Fig. 2). Chestnuts are ring-porous trees and must make new xylem vessels every spring, as the previous year's vessels become occluded. If the cambium is killed, and no new vessels are formed, buds will break in the spring but soon wilt and die. Everything distal to the "canker" then dies. When apical dominance is removed, epicormic sprouts are formed along the stem. Sprouts also form from a burl-like tissue at the base of the tree called the "root collar", which contains dormant embryos. Sprouts grow, become wounded and infected, and die, and the process starts all over again.

Cankers were first reported in 1904 on American chestnut trees lining the avenues of the Bronx Zoo in New York City (Anagnostakis, 1987, 1993). Attempts were made to control the chestnut blight epidemic (chemical treatments, clearing and burning chestnut trees around infection sites) but nothing worked. By 1950 the fungus was reported throughout the native range of *C. dentata*; from Maine to Georgia, and west to the edge of Michigan (Fig. 3). A major forest tree had been reduced to a multiple-stemmed shrub. In 1912 the Plant Quarantine Act was passed to reduce the chances of such

Fig. 2.   Canker on an American chestnut stem caused by *Cryphonectria parasitica*. The fungus probably entered the bark at the broken branch. Growth in and under the bark is roughly circular, and stromata have broken through the lenticels. Photographed by R. A. Jaynes, The Connecticut Agricultural Experiment Station, New Haven, Connecticut.

a catastrophe happening again (Waterworth and White, 1982).

When Haven Metcalf and J. Franklin Collins wrote their 1909 Bulletin they stated:

Even [in 1904] it is certain that [chestnut blight] had spread over Nassau County and Greater New York, and had found lodgment in the adjacent counties of Connecticut and New Jersey. No earlier observation than this is recorded, but it is evident that the disease, which would of necessity have made slow advance at first, must have been in this general locality for a number of years in order to have gained such a foothold by 1904. Conspicuous as it is, it is strange that the fungus causing this disease was not observed or collected by any mycologist until May, 1905, when specimens were received from New Jersey by Mrs. F. W. Patterson, the Mycologist of the Bureau of Plant Industry. . . . By August, 1907, specimens received by this Bureau showed that the disease had reached at least as far south as Trenton, N.J., and as far north as Poughkeepsie, N.Y., and was spread generally over Westchester and Nassau Counties, N.Y., Bergen County, N.J., and Fairfield County, Conn. . . . reports have been received from points as remote as Cape Cod, Wellesley and Pittsfield, Mass.; Rochester and Shelter Island, N.Y., and Akron, Ohio. . . . It can be quite confidently stated that the bark disease does not yet occur south of Virginia . . .

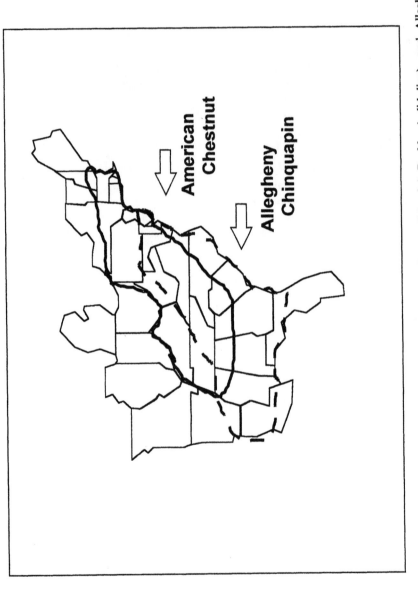

Fig. 3. Native range of American chestnut trees, *Castanea dentata* (Marsh.) Borkh. (solid line), and Allegheny chinquapin (dashed line). Modified from Johnson (1988).

The theory ... that the Japanese chestnuts were the original source of infection, has been strengthened by many facts. ...

While the disease has spread principally from the vicinity of New York [City] there is much to indicate that it occurred at other points at an early date. Chester's *Cytospora* on a Japanese chestnut noted at Newark, Del., in 1902, may have been the bark disease. Observations by the junior writer indicate that this disease may have been present in an orchard in Bedford County, Va. as early as 1903, and that in Lancaster County, Pa., it probably was present as early as 1905.

It becomes more and more evident as this disease is studied that diseased nursery stock is the most important factor in its spread to distant points.

In 1913, David Fairchild of the USDA asked plant explorer Frank Meyer to look for the disease in Asia. Meyer reported in early June that he had found it in China. He wrote:

This blight does not by far do as much damage to the Chinese chestnut trees as to the American ones. Not a single tree could be found which had been killed entirely by this disease, although there might have been such trees which had been removed by the ever active and economic Chinese farmers.

Shear and Stevens grew cultures from Meyer's samples, and in July they inoculated the Chinese fungus into American trees near Washington, DC. Rapid death of the sprouts confirmed that this similar-appearing fungus caused chestnut blight.

Meyer went to Japan in 1915 and was again first in finding chestnut blight. He wrote that the Japanese chestnut trees were generally more resistant to the blight disease than the Chinese chestnut trees that he had seen.

The blight fungus has maintained its destructive vigor for at least 80 years in New England. Sprouts from the root collar perpetuate the species (Paillet, 1993), but these are usually killed before they become sexually mature. Since two flowering chestnut trees are needed for seed formation (they must be cross-pollinated), sexual reproduction has been drastically reduced and little resistance to the pathogen has evolved. In the full sun of a clear-cut the cycle of sprouting–infection–death–sprouting takes about ten years, and in the shady understory of a forest the cycle may take 30 years. In spite of this inevitable lethality, American chestnut root sprouts are still a major component of subcanopy and shrub biomass (McCormick and Platt, 1980; Paillet, 1982; Adams and Stephenson, 1983; Stephenson *et al.*, 1991).

A glimmer of hope was seen in the early 1960s when Jene Grente, of France, isolated abnormal strains of the chestnut blight fungus from chestnut trees north of Milan, near Bergamo, Italy. Grente reported that these "hypovirulent" strains lacked the ability to kill chestnut trees, and that inoculation of these strains into existing (lethal) cankers resulted in canker remission (Grente, 1965). Grente published several papers describing hypovirulence, and began a program of active intervention when blight was found in France. Hypovirulent (H) strains were sent to extension agents who used them to treat cankers in French chestnut orchards. Treatment of new cankers as they formed resulted in a successful "biological therapy" of the disease. After four or five years of therapy, H strains began to spread through the chestnut orchards of France, the trees began "healing" over the blight cankers with lumpy, bark-callus tissue (as was happening in Italy), and a biological control was established (Grente and Berthelay-Sauret, 1978).

Many European H strains, such as those originally described by Grente, lack pigment, and do not have the same morphology and growth rate as virulent strains in culture in the laboratory. The biological control of chestnut blight in the orchards of C. *sativa* in France and Italy spurred searches in the USA for H strains. Similar "healing" cankers in Michigan, Tennessee, Virginia, and West Virginia yielded orange strains of the blight fungus that were also less able to kill chestnut trees (Day *et al.*, 1977).

Grente knew that his H strains had some kind of cytoplasmic "disease", but did not know the cause. A survey of fungi for viruses (Moffitt and Lister, 1975) reported that two of Grente's hypovirulent strains reacted to antibodies made to double-stranded (ds)RNA, the genetic material of most fungal viruses, but they could not detect dsRNA or virus-like-particles in extracts. In 1972, Grente sent H cultures to The Connecticut Agricultural Experiment Station, where it was confirmed that Grente's H strains, and the US Hypovirulent types, contained dsRNA (Day *et al.*, 1977), and could control American, virulent strains of *C. parasitica* on American chestnut trees (Anagnostakis and Jaynes, 1973). When mycelium of an H strain is placed into holes in the bark around killing cankers, viruses can move into the virulent strains that caused the cankers. The cankers then stop expanding, and the tree's natural defenses of walling off invaders succeeds in protecting the cambium.

Virologists now consider that the European and American H strains of *C. parasitica* are infected with viruses of three very different types (three "families"), with wide variation in their effects on the fungus, and with varied effectiveness for biological control of chestnut blight (Anagnostakis and Hillman, 1992). All have dsRNA as their genetic material, have no capsid (but are surrounded by fungal-host membrane), and resemble POTY viruses. There are four open reading frames, and two of the viruses have been sequenced (Nuss, 1992). The current taxonomy places most of the viruses studied (including the European types) in the family Hypovirideae (Hillman *et al.*, 1994).

It was soon discovered that there are barriers to virus transmission. Vegetative compatibility (v-c) genes in the fungus limit the number and stability of fusion bridges between strains (Fig. 4), and there are many strains present in areas where the fungus has been present for a long time with an opportunity to diversify (Anagnostakis, 1992a). Virus transmission will always occur if opposing strains have the same alleles at all of their v-c genes (estimated to be seven or eight loci), and transmission can occur when few, or certain, loci have different alleles. Mixtures of hypovirulent strains, representing the most common v-c types in the plot, can be used for treatment, increasing conversion efficiency (Jaynes and Elliston, 1980; Turchetti and Maresi, 1988; Anagnostakis, 1990).

When we had the permission of the US Plant Quarantine Division (in 1973), tests were begun on trees growing outside at The Experiment Station Farm (Lockwood Farm) in Hamden, Connecticut. In 1976 an orchard was planted at Lockwood Farm with 71, three-year-old American chestnut seedlings raised by E. Thor, of the University of Tennessee. The seed for these came from Michigan and Wisconsin, where blight had not yet reached the *C. dentata* planted outside the native range. Our trees became infected with the chestnut blight fungus, and were treated for four consecutive years (from 1978 through 1981) with a mixture of H strains. In 1981, 43 of the original, main stems were still alive (J. E. Elliston, unpublished). In 1992, 28 of the original stems survived, and there were 28 additional sprout-clumps with stems that reached the

Fig. 4. A test of vegetative compatibility (v-c) with *Cryphonectria parasitica*. An unknown strain (left member of each pair) paired with the standard testers for v-c groups 1 to 10. The medium was Difco potato dextrose agar, and the plate was incubated at 28°C in the dark. The unknown strain has merged with the v-c 2 tester strain (second down on the left) and so is v-c type 2. It has formed a barrage line with the nine other tester strains.

canopy. New cankers form each season, but usually become hypovirulent, and are restricted by host response to superficial layers of phloem, before much cambium has been killed.

Some of these individual trees seem to be surviving better than others. However, since we have no records of the exact source of each of the trees, it is not possible to draw conclusions about population fitness. Perhaps some are simply better able to thrive with extensive hypovirulent cankers. Nut production is heavy, and since 1989 these trees have been used in a back-cross breeding program. When cankers in this plot kill twigs or branches, they are cut away (pruning is done in July to minimize new infections). We assume that animals, birds, insects, and gastropods are moving hypovirulent strains of the fungus from cured cankers to newly formed killing cankers.

C. parasitica spores formed on the bark surface can easily be picked up on the feet and fur of animals, or feet and feathers of the many birds that frequent chestnut trees (Sharf and DePalma, 1981). In addition to these highly visible creatures, there are many small visitors to chestnut blight cankers. Mites of many species have been found in chestnut blight cankers (Wendt *et al.* 1983; Nannelli and Turchetti, 1989), and everyone who has had a mite-infestation of their culture incubator knows their potential for transporting spores. Other insects that frequent chestnut trees are carpenter worms (*Prionoxystus robiniae*), and clear-winged moths (*Synanthedon castaneae*) that presumably lay their eggs only in American chestnut bark. The latter were thought to be extinct until traps with pheromones revealed their presence in Georgia and in our orchard at Lockwood Farm (Anagnostakis *et al.*, 1994).

Other potential H-strain carriers include tree-climbing slugs. Turchetti and Chelazzi (1984) found *Lehmannia marginata* moving and feeding on chestnut blight cankers in the woods in Italy, and both virulent and H strains were cultured from fecal pellets of this species. Since the slug *Philomycus carolinianus* was repeatedly found on chestnut trees in Connecticut, some effort was made to study these fungivores. We were unable to prove that they carried *C. parasitica*, but this question is worthy of examination in greater detail.

Among the most obvious candidates for carriers of H strains are carpenter ants. Their main nesting sites are standing dead trees or poles, and they can forage 30–35 meters from their nests. Their principal food is honey dew from aphids, but they also eat fungi. They can store food in their crops for extended periods, and then regurgitate their crop contents to feed comrades. *Camponotus pennsylvanicus* and *C. p. ferruginia* are frequently seen on chestnut trees in Connecticut (the latter are the most common). They tend colonies of the leaf aphids *Colaphis castaneae*, and stem aphids *Petchia virginiana*. In experiments on chestnut trees in a wood lot *C.p. ferruginia* were frequently observed chewing on newly initiated blight cankers, but never chewing on fungus-free agar medium (potato dextrose agar) in holes in the bark, or on the non-infected bark itself. We have collected aphid guardians and foragers of both carpenter ant species, and recovered live blight fungus propagules from their crops (Anagnostakis, 1982).

Our first tests of biocontrol of chestnut blight in the forest were mapped areas (circles, 25 m radius) with American chestnut trees that were treated with mixtures of H strains for four years (1978–1981), and surrounding circles (from 25 m out to 50 m) mapped but not treated. R. A. Jaynes and N. DePalma set up the four plots, and reported that by 1981 treated trees had larger average diameters at breast height (dbh) and higher percentage survival than non-treated trees in the surrounding annulus (5.7 and 4.8 cm dbh, 81% v.s. 58% survival). In 1983, treated stem survival was 27%, vs. 15% for the non-treated trees, and average stem dbh was about 7 and 4.5 cm (Jaynes and DePalma, 1984). In 1987 I made transects through two of the plots, 100 m out from the centers, and recorded information on stems >2.5 cm dbh. I found more live stems in the central, treated areas than in nearby plots where no H strains had been introduced, but Jaynes and DePalma's non-treated, outer donuts had about as many surviving chestnut stems as the treated, inner circles. The average dbh of stems in the center was 6 cm, non-treated stems 30–50 m from the center averaged about 5 cm, and the stems 51–100 m from the center averaged about 3 cm. I found H cankers as far as 75 m from the edges of the treated circles. There was a lot of competition from other woody plants in all of the plots, and many of the original, treated stems were dead. Griffin (1992) showed that chestnut survival is heavily dependent on the amount of competition the trees have from other plant species. Despite the mortality, the data clearly showed that hypovirulence could be maintained and spread in chestnut blight fungus populations in the forest, thus improving tree survival (Anagnostakis, 1990).

A forest plot in Rocky Hill, Connecticut, where the C. parasitica population has been monitored for many years has given similar results. From 1982 through 1986, the 317 cankers that developed on American chestnut stems in 50 sprout clumps were treated with H strains. By 1991, there were 484 cankers on these sprouts. The number of v-c types increased from six, in 1982, to 48 in 1985, and we stopped counting (Anagnostakis and Kranz, 1987). In 1988 and 1993 we measured all of the stems at least 2.5 cm dbh in the treated sprout clumps, and in 60 other clumps, to the south on the same property, for controls. In 1988, the average dbh of treated stems was 7.3 cm and of control stems was 6.0 cm. In 1993, treated stems averaged 8.1 cm and controls 4.4 cm dbh. Even more indicative of the difference between treated and control stems was the fact that in 1988, 16% of the treated stems and 12% of the control stems were at least 10 cm dbh. In 1993, 37% of the treated stems, but only 3% of the control stems were at least 10 cm dbh.

It is clear that hypovirulence is not succeeding as well at controlling chestnut blight in West Virginia as it is in Connecticut (W. L. MacDonald, West Virginia University, Morgantown, personal communication). There are several possible explanations for this. First, the trees may be different in each population (provenance) in the US. Second, there are many different kinds of viruses in the blight fungus population in West Virginia. We have never found "native" viruses in our *C. parasitica* isolates from Connecticut. If these interfere with the infection of the fungus by hypovirulence viruses, it might reduce the possibility for biological control. The third "southern-factor" is the presence, in states from Pennsylvania south, of a fungus that is closely related to the blight fungus, and has never been found in Connecticut. *Cryphonectria (fluens) radicalis* has not been found in the US since chestnut blight reached its native range. The two fungi might be interacting (or might have interacted) in some way that has not yet been discovered. It is also possible that *Phytophthora cinnamomi* is involved in this very complex scenario. It is a challenge for us to find reasons for the differences in usefulness of hypovirulence.

Hypovirulence can be used to help us grow American or hybrid chestnut trees in some parts of the country, but is not good enough (yet) to allow us to grow a useful timber-crop of American chestnut trees. It is likely that one problem is the 10%–50% of the conidia produced by hypovirulent strains that do not contain the virus (and thus are fully virulent), and the ascospores (100%) that do not contain the virus. A possibility for overcoming this dilemma has been provided by D. L. Nuss and his colleagues at the Roche Institute in New Jersey. They have made a cDNA copy of the genome of our most frequently used French virus, and transformed it onto a chromosome of *C. parasitica* (Choi and Nuss, 1992). These recombinant strains are stable in culture. All conidia contain the viral genes and are hypovirulent, and half of the ascospore progeny, resulting from using recombinant strains as males in crosses, contain the viral genes. In addition, these strains reproduce the viral genome as dsRNA in the cytoplasm, so that transmission by hyphal fusion also occurs. Deletion or change of specific sequences of the cDNA will provide clues to the action of the genome in preventing virulence in *C. parasitica*, and perhaps ultimately to the nature of virulence itself. Greenhouse tests of the pathogenicity of a recombinant strain on other woody plant species have been done, and permission has now been obtained from the federal government (USDA/APHIS/BBEP) to use these strains on chestnut trees in the forests. I have set up plots in Connecticut, and W. L. MacDonald has similar plots in West Virginia, so that both northern and mid-Atlantic American chestnut trees (in northern and mid-Atlantic ecosystems) can be tested for their response to biological control of chestnut blight by the new hypovirulent strains.

While the nature of hypovirulence is being explored and improved, this biological control can be used to keep orchards of American chestnut trees alive. This allows us to preserve the diversity of the tree in provenances throughout its range, and allows us to go back to breeding for resistance to chestnut blight and ink disease, and for improvements in nut size and quality.

Hypovirulence will also be useful if partially resistant trees are chosen as good nut producers, and planted in commercial orchards.

One of the results of the chestnut blight epidemic was a massive effort by the US government and many private individuals to find trees to replace American chestnuts in the forests. The species of the genus *Castanea* are listed in Table 1. What chestnut germplasm is available to us in the United States, and how similar is the stock to American chestnut?

IV. CHESTNUT GERMPLASM

A. EUROPEAN CHESTNUT TREES

European chestnuts (*C. sativa*) were distributed throughout southern Europe from the Caucasus mountains, and the nuts have become an important food source, both cooked whole and ground into flour. The wild trees are similar in stature to American chestnuts, but their maximum height is about 25 m whereas *C. dentata* may be 30 m. Most *C. sativa* sold commercially have been selected for nut production, and are shorter and more spreading than *C. dentata*. There are two recorded, early importations of this species into the US. First, Thomas Jefferson brought cuttings to his home, Monticello, in 1773 and grafted them onto native American chestnut trees. Second, Eleuthere Irenee DuPont de Nemours, who in 1799 moved from France to Bergen Point, New Jersey, and then to Brandywine, Delaware, brought many European chestnuts with him, imported more later, crossed them with native trees, and planted them all over the area. This probably contributed to the diversity of chestnuts in the south-eastern part of the US, but *C. sativa* is also highly susceptible to chestnut blight and ink disease.

B. JAPANESE CHESTNUT TREES

G. H. Powell wrote in 1900 that Japanese chestnut trees (*C. crenata*) were first imported in 1876 by nurseryman S. B. Parsons, of Flushing, NY (in the New York City borough of Queens, at the western end of Long Island). These were widely distributed, and two of them were planted and still survive in southern Connecticut. Their blight resistance is usually good, but trees of this species are rarely more than 15 m tall, and have low, spreading forms that make them well suited to orchards, but not usable for timber. In 1882, William Parry, in New Jersey, imported 1000 grafted Japanese chestnut trees. In the West, Luther Burbank planted a box of seeds sent by his collector from Japan in 1886. He subsequently had more than 10 000 bearing trees growing in his Santa Rosa, California, nursery, and in 1893 his "New Creations" catalog advertised a "New Japan Mammoth" chestnut.

There were 21 varieties of Japanese chestnuts listed in T. H. Powell's 1898

Bulletin. These were discussed in gardening magazines such as *The Rural New Yorker*, and advertized in plant and seed catalogs. Mail order spread *C. crenata* all over the country (Table III). Powell also reported that (by 1899)

TABLE III

Nurseries in the US selling chestnut trees by mail-order, around the turn of the century[a]

Catalog	Date	Species, "cultivar"	Cost each ($)
Reading Nursery, Jacob W. Manning, Reading, Massachusetts	1900	American	0.50–1.00
J. T. Lovett Co., Little Silver, New Jersey	1888	"Japan Giant"	0.75
		Spanish	0.30
		American	0.10–0.25
		"Numbo"	0.75
Storrs and Harrison, Painesville, Ohio	1888	American	0.50
		"Japan Giant"	0.50–0.75
		Spanish	0.50
Shady Hill Nursery, Cambridge (Somerville), Massachusetts	1888/1889	American	0.10–0.35
Highlands Nursery, H. P. Kelsey, Boston, Massachusetts	1899/1900	American	0.25
Biltmore Nursery, Biltmore, North Carolina	1900/1901	American	0.15–0.50
Mt. Hope Nursery, Ellwanger and Barry, Rochester, New York	1897	C. Americana	0.50
		C. Japonica	1.00
		C. vesca	0.50
Elm City Nursery, New Haven, Connecticut	1901	American	0.50–1.00
		Spanish	0.25–1.00
		"Numbo"	1.50
		Japanese	0.50–1.00
Fruitland Nurseries, P. J. Berckmans, Augusta, Georgia	1900	American	0.25–1.00
		Spanish	0.25
J. H. Hale, South Glastonbury, Connecticut	1903	"Coe", "Hale" "McFarland"	
C. B. Hornor and Son, Mt. Holly, New Jersey	1897	American	0.25–0.35
		"Numbo"	0.75
		"Paragon"	1.00–2.50

[a] This information was compiled by the author from catalogs in the collections of the New York Botanical Garden, in New York City, and the Arnold Arboretum, in Jamaica Plain, Massachusetts.

there were over 300 acres of chestnut trees near Philadelphia grafted with European and Japanese varieties, and that the Lovett Co., in Little Silver, New Jersey (near the coast, about 15 miles south of Long Island) had also imported Japanese chestnut trees and was selling them by mail-order (Powell, 1900). In 1930 when Arthur Graves was looking for resistant trees for breeding he found large Japanese chestnut trees on several estates on Long Island (New York) and in northern New Jersey. He said that many of them had been purchased around the turn of the century as "Japanese Giant" from a nursery near Rochester.

C. CHINESE CHESTNUT TREES

In 1908, E. H. Wilson sent The Arnold Arboretum (near Boston, Massachusetts), seeds of his collection #551, *C. Henryi* from Western Hupeh, China. This species is reported to be a timber tree in China, but the only known survivor in the US (in Connecticut) is only 8 m tall, and quite sensitive to winter damage. Another Chinese chestnut, *C. seguinii*, is a dwarf species that has not been widely planted here.

In 1903 Dr Charles Sprague Sargent sent *C. mollissima* seed to the US, and The Plant Introduction Department in the US Bureau of Plant Industry started importing *C. mollissima* seed and scions in large numbers in 1912. This species, called Chinese chestnut, is usually quite resistant to chestnut blight, and it is now the most common Asian chestnut growing in the US. The timber-type trees of *C. mollissima* may be 20 m tall, and could have limited use for poles. However, most of the seeds imported have been of nut-producing varieties that have spreading shapes and early, heavy fruiting — characteristics important in orchard trees but not suitable for timber trees.

D. HYBRIDIZATION

Meyer suggested in 1915 that "This Japanese chestnut, *Castanea japonica* might be used as a factor in hybridization experiments together with American, European, and Chinese species to create immune or nearly immune strains of chestnuts."

Many people took up Meyer's suggestion, and hybrids made earlier to improve the orchard qualities of chestnut trees were examined for their resistance to chestnut blight. Chestnuts are actually fruits, with shells enclosing cotyledons. Trees bear both male and female flowers in late spring, but must be cross-pollinated for nut production. Nuts are born in a spiny involucre, or bur, which opens to release the nuts in late fall. Our first records of crosses between chestnut species typify the whole history of chestnut breeding in the

US: the work was done by both an interested amateur and by a professional botanist (Anagnostakis, 1992b). George W. Endicott of Villa Ridge, Illinois was growing "Japan Giant" at the end of the last century, and used pollen from an American chestnut tree to produce Japanese × American hybrids in 1895. The other early hybridization work was done by Dr Walter Van Fleet, then an associate editor of the *Rural New Yorker Magazine*. In 1894 he used pollen of American chestnut on flowers of the European (or European–American) cultivar "Paragon" and planted the progeny in Little Silver, New Jersey. Van Fleet went on to make thousands of crosses, using many species, between 1900 and 1921. His early crosses used the native chinquapin, *C. pumila*, and European and Japanese cultivars. In his later work he included the Chinese chestnut, *C. mollissima*. Van Fleet had over 900 of these trees to observe and use at the USDA station in Bell, Maryland. The only hybrid of Van Fleet's that survives today is a seedling of his S-8 (*C. crenata × C. pumila*), probably open-pollinated by a Japanese chestnut.

The longest-continuing chestnut breeding program in the United States is that in Connecticut. Dr Arthur H. Graves planted trees on land that he owned in Hamden, Connecticut, and started making crosses in 1930. Two of his students, Hans Nienstaedt and Richard A. Jaynes, maintained trees, made crosses, and contributed greatly to our knowledge of chestnut in general. In 1947 Graves deeded his land with the Sleeping Giant Chestnut Plantation to the State, to ensure that the work would continue. Since then the Plantation has been maintained by The Connecticut Agricultural Experiment Station, and is probably the finest collection of species and hybrids in the world.

The early Connecticut breeding work focused on making hybrids that were combinations of species, looking for single ideal progeny that could be propagated clonally. Jaynes cooperated with the Virginia Division of Forestry to plant over 10 000 hybrid chestnut seedlings in the Lesesne State Forest, in Virginia. These are still being observed by T. Dieroff of the Virginia Division of Forestry, so that promising trees can be selected.

With Jaynes's retirement in 1984, responsibility for the chestnut breeding program fell to me. At the urging of geneticist Charles R. Burnham, records were searched for hybrids that were products of resistant × susceptible trees, and any that were backcrossed again to the susceptible parent species. Burnham felt that a few generations of backcrossing and selecting, followed by crosses of those products, would allow selection of trees that had the form and nut quality needed, combined with resistance to chestnut blight (Burnham, 1988). These trees would produce "true to type" offspring, and allow reforestation with chestnut.

Because of the long-term commitment of The Connecticut Agricultural Experiment Station in maintaining records and valuable trees, this project is proceeding. Many Asian trees in Connecticut have been evaluated for survival in our climate, resistance to chestnut blight, timber form, and nut quality. Selected trees have been used to make new hybrids with American chestnut trees kept alive using biological control by hypovirulence. Experiment Station trees are also being used by F. V. Hebard, chestnut breeder for the American Chestnut Foundation, and progeny are planted at the Wagner Research Farm in Virginia. Two first-generation-backcross (BC1) trees [(Chinese × American) × American] are now 48 and 41 years old. The first is a hybrid made by Diller in 1946, called "Clapper". Although the original tree has died, two grafts survive at our Lockwood Farm. Our other old BC1 tree was made by Graves and Nienstaedt in 1953, and we call it "Graves". Both have timber form, good blight resistance, and acceptable nuts. Hebard has inoculated the chestnut blight fungus into 438 seedlings of "Graves" × "Clapper" and preliminary results suggest that 28 are fully resistant (F. V. Hebard, American Chestnut Foundation, Meadowview, Virginia, personal communication). If this ratio holds true, there probably cannot be more than two genes for resistance to chestnut blight in Chinese chestnut trees, and the backcross breeding program has a good chance of succeeding.

Selections are now being made for orchard as well as timber trees. Some of the complex hybrids made by Jaynes have the Chinese shrub *C. seguinii* in their background, and are compact dwarfs. These have been used as dwarfing rootstocks (to get early flowers on short trees), and are now being used in crosses with chestnut trees with exceptional nut quality to select short, reliable nut producers.

V. CONCLUSION

There are still further problems. Two species of weevils lay eggs in the nuts, and the developing larvae make the nuts unmarketable (Payne and Johnson, 1979). The larvae of the clear-winged moth, *Synanthedon castaneae*, tunnel in bark and wood causing enough damage that they may soon be a cause for concern, especially if they become as prevalent as their relatives the dogwood and peach borers. We will have to find a way to protect our trees from this insect.

The last, and probably most serious, pest threatening our chestnut trees is the oriental chestnut gall wasp, *Dryocosmus kuriphilus*. This insect was brought into Georgia in 1974 from Asia, on cuttings that did not go through plant quarantine (Payne *et al.*, 1976). Chinese chestnut trees that were grown in Georgia orchards for their nut crop, were soon infested. The insect lays eggs in the vegetative and mixed buds and, as the larvae develop, green or rose-colored galls are formed on the leaves, petioles, and even catkins. These galls suppress shoot elongation and reduce fruiting: trees with severe infestations lose their vigor and often die. American chestnut seedlings from Southington, Connecticut, were planted in Byron, Georgia, in 1989, and by 1992 these trees were heavily galled (Fig. 5). In June 1993 galls of *D. kuriphilus* were found on American chestnut trees growing along the Appalachian Trail in the

Fig. 5.   American chestnut (*Castanea dentata*) infested with the gall-causing wasp *Dryocosmus kuriphilus*. Photographed by J. A. Payne, USDA/ARS, Fruit and Tree Nut Research Laboratory, Byron, Georgia.

Chattahoochee National Forest in northern Georgia, at the southern end of the native range of this tree (Anagnostakis and Payne, 1994), and the insect is now also in southern North Carolina and southern Tennessee.

Spread of the gall wasp occurs as a result of movement of infested twigs or shoots, or by flight of the adults. Since the Appalachian Trail extends from northern Georgia to Maine, and since American chestnuts are found along the whole length of the trail, there will be a source of host material for the gall wasp to allow it to move throughout the native range of our tree. J. A. Payne (USDA/ARS, Fruit and Tree Nut Research Laboratory, Byron, Georgia) has observed that chinquapins seem to be resistant or immune to this insect, and

I have started crossing our various chestnut-breeding lines with chinquapins. Progeny are being planted in Georgia and Tennessee to select for those resistant to Gall Wasp and chestnut blight (and maybe even ink disease), and I am hoping to have resistant trees before the insect reaches Connecticut.

The work at The Connecticut Experiment Station will continue, and the renewed interest in chestnuts should allow cooperation with many people to speed progress towards usable chestnut timber stands and a new nut market in the United States. The basic principles established will help us to study other plant diseases and their control, improve plant breeding programs, and perhaps even convince people that plant quarantine is very important.

REFERENCES

Abreu, C. A. (1992). Chestnut ink disease: management practices and resistance. *In* "Proceedings of the World Chestnut Industry Conference" (R. E. Wallace and L. G. Spinella, eds) pp. 153–157. Chestnut Marketing Association, Alachua, Florida.

Adams, S. M. and Stephenson, S. L. (1983). A description of the vegetation on the south slopes of Peters Mountain, southwestern Virginia. *Bulletin of the Torrey Botanical Club* 110, 18–23.

Anagnostakis, S. L. (1982). Carpenter ants as carriers of *Endothia parasitica*. *In* "Proceedings, USDA Forest Service American Chestnut Cooperators' Meeting" (H. C. Smith and W. L. MacDonald, eds), pp. 111–113. West Virginia University Press, Morgantown.

Anagnostakis, S. L. (1987). Chestnut blight: the classical problem of an introduced pathogen. *Mycologia* 79, 23–37.

Anagnostakis, S. L. (1990). Improved chestnut tree condition maintained in two Connecticut plots after treatments with hypovirulent strains of the chestnut blight fungus. *Forest Science* 36, 113–124.

Anagnostakis, S. L. (1992a). Diversity within populations of fungal pathogens on perennial parts of perennial plants. *In* "The Fungal Community: Its Organization and Role in the Ecosystem", 2nd edition (G. C. Carroll and D. T. Wicklow, eds), pp. 183–192. Marcel Dekker, New York.

Anagnostakis, S. L. (1992b). Chestnut breeding in the United States. *In* "Proceedings of the World Chestnut Industry Conference" (R. E. Wallace and L. G. Spinella, eds), pp. 19–21. Chestnut Marketing Association, Alachua, Florida.

Anagnostakis, S. L. (1993). Chestnuts and the introduction of chestnut blight. *Annual Report of the Northern Nut Growers Association* 83, 39–42.

Anagnostakis, S. L. and Hillman, B. (1992). Evolution of the chestnut tree and its blight. *Arnoldia* 52, 3–10.

Anagnostakis, S. L. and Jaynes, R. A. (1973). Chestnut blight control: use of hypovirulent cultures. *Plant Disease Reporter* 57, 225–226.

Anagnostakis, S. L. and Kranz, J. (1987). Population dynamics of *Cryphonectria parasitica* in a mixed-hardwood forest in Connecticut. *Phytopathology* 77, 751–754.

Anagnostakis, S. L. and Payne, J. A. (1094). Oriental chestnut gall wasp found in American chestnut trees. *Entomology Bulletin Board*

Anagnostakis, S. L., Welch, K. M., Snow, J. W., Scarborough, K. and Eichlin, T. D. (1994). The rediscovery of the clearwing chestnut moth, *Synanthedon castaneae*

(Busck) (Lepidoptera: Sesiidae) in Connecticut. *Journal of the New York Entomological Society* **102**, 188–189.

Breisch, H. (1992). Compatibility tests between the main French varieties of chestnut trees and ink-resistant hybrid rootstocks. *In* "Proceedings of the World Chestnut Industry Conference" (R. E. Wallace and L. G. Spinella, eds), pp. 41–47. Chestnut Marketing Association, Alachua, Florida.

Burnham, C. R. (1988). The restoration of the American chestnut. *American Scientist* **76**, 478–487.

Choi, G. H. and Nuss, D. L. (1992). Hypovirulence of chestnut blight fungus conferred by an infectious viral cDNA. *Science* **257**, 800–803.

Crandall, B. S., Gravatt, G. F. and Ryan, M. M. (1945). Root disease of *Castanea* species and some coniferous and broadleaf nursery stocks, caused by *Phytophthora cinnamomi*. *Phytopathology* **35**,162–180.

Day, P. R., Dodds, J. A., Elliston, J. E., Jaynes, R. A. and Anagnostakis, S. L. (1977). Double-stranded RNA in *Endothia parasitica*. *Phytopathology* **67**, 1393–1396.

Grente, J. (1965). Les formes Hypovirulentes d'*Endothia parasitica* et les espoirs de lutte contre le chancre du châtaignier. *Académie d'Agriculture de France, Extrait du Proces-verbal de la Séance* **51**, 1033–1037.

Grente, J. and Berthelay-Sauret, S. (1978). Biological control of chestnut blight in France. *In* "Proceedings of the American Chestnut Symposium" (W. L. MacDonald, F. C. Cech, J. Luchock, and C. Smith, eds) pp. 30–34. West Virginia University, Morgantown.

Griffin, G. J. (1992). American chestnut survival in understory mesic sites following the chestnut blight pandemic. *Canadian Journal of Botany* **70**, 1950–1956.

Hillman, B. I., Fulbright, D. W., Nuss, D. L. and Van Alfen, N. K. (1995). Hypoviridae. *In* "Virus Taxonomy: Sixth Report of the International Committee for the Taxonomy of Viruses" (F. A. Murphy, C. M. Fauquet, D. H. L. Bishop, S. A. Ghabrial, A. W. Jarvis, G. P. Martelli, M. P. Mayo, and M. D. Summers, eds). Springer-Verlag, Wein, New York, p. 264.

Jaynes, R. A. and DePalma, N. K. (1984). Effects of blight curing strains on native chestnut for 6 years. *Annual Report of the Northern Nut Growers Association* **75**, 64–71.

Jaynes, R. A. and Elliston, J. E. (1980). Pathogenicity and canker control by mixtures of hypovirulent strains of *Endothia parasitica* in American chestnut. *Phytopathology* **70**, 453–456.

Johnson, G. P. (1988) Revison of *Castanea* sect. Balanocastanon (Fagaceae). *Journal of the Arnold Arboretum* **69**, 25–49.

McCormick, J. E. and Platt, R. B. (1980). Recovery of an Appalachian forest following the chestnut blight. *American Midland Naturalist* **104**, 264–273.

Metcalf, H. and Collins, J. F. (1909). The present status of the chestnut bark disease. *USDA Bureau of Plant Industry Bulletin* #141, part V.

Moffitt, E. M. and Lister, R. M. (1975). Application of a serological screening test for detecting double-stranded RNA mycoviruses. *Phytopathology* **65**, 851–859.

Nannelli, R. and Turchetti, T. (1989). Osservazioni preliminari sul'associazione di alcune specie di acari corticicole con *Cryphonectria parasitica* (Murr.) Barr. *Redia* **72**, 581–593.

Nuss, D. L. (1992). Biological control of chestnut blight: an example of virus-mediated attenuation of fungal pathogenesis. *Microbiological Reviews* **56**, 561–576.

Paillet, F. L. (1982). Ecological significance of American chestnut in the Holocene forests of Connecticut. *Bulletin of the Torrey Botanical Club* **109**, 457–473.

Paillet, F. L. (1993). Growth form and life histories of American chestnut and

Allegheny and Ozark chinquapin at various North American sites. *Bulletin of the Torrey Botanical Club* **120**, 257–268.

Payne, J. A., Green, R. A. and Lester, D. D. (1976). New nut pest: an oriental chestnut gall wasp in North America. *Annual Report of the Northern Nut Growers Association.* **67**, 83–86.

Payne, J. A. and Johnson, W. T. (1979). Plant pests. *In* "Nut Tree Culture in North America" (R. A. Jaynes, ed.), pp. 314–364. Northern Nut Growers Association, Inc.

Powell, G. H. (1898). The European and Japanese chestnuts in the Eastern United States. Delaware College Agricultural Experiment Station Bulletin #42, Newark, Delaware.

Powell, G. H. (1900). The European and Japanese chestnuts in the Eastern United States. *In* 11th Annual Report of the Delaware College Agricultural Experiment Station, pp. 101–135. Newark, Delaware.

Sharf, S. S. and DePalma, N. K. (1981). Birds and mammals as vectors of the chestnut blight fungus (*Endothia parasitica*). *Canadian Journal of Zoology* **59**, 1647–1650.

Stephenson, S. L., Adams, H. S. and Lipford, M. (1991). The present distribution of chestnut in the upland forest communities of Virginia. *Bulletin of the Torrey Botanical Club* **118**, 24–32.

Turchetti, T. and Chelazzi, G. (1984). Possible role of slugs as vectors of the chestnut blight fungus. *European Journal of Forest Pathology* **14**, 125–127.

Turchetti, T. and Maresi, G. (1988). Mixed inoculum for the biological control of chestnut blight. *European and Mediterranean Plant Protection Organization Bulletin* **18**, 67–72.

Waterworth, H. E. and White, G. A. (1982). Plant introductions and quarantine: the need for both. *Plant Disease* **66**, 87–90.

Wendt, R., Weidhaas, J., Griffin, G. J. and Elkins, J. R. (1983). Association of *Endothia parasitica* with mites isolated from cankers on American chestnut trees. *Plant Disease* **67**, 757–758.

Fungal Avirulence Genes and Plant Resistance Genes: Unraveling the Molecular Basis of Gene-for-gene Interactions

PIERRE J. G. M. DE WIT

Department of Phytopathology, Wageningen Agricultural University, PO Box 8025, 6700 EE Wageningen, The Netherlands

Advances in Botanical Research Vol. 21
incorporating Advances in Plant Pathology
ISBN 0-12-005921-5

1. INTRODUCTION

The phenomenon of resistance of plants to fungal pathogens was first described by Farrer nearly a century ago (Farrer, 1898). He showed that certain wheat cultivars were resistant to *Puccinia graminis* f. sp. *tritici*. Biffen (1905) proved that resistance of wheat to yellow rust was genetically determined and followed Mendel's laws. However, within 10 years after introduction of rust resistance genes by breeding it was found that stem rust-resistant cultivars of wheat became diseased by new biologic forms which later became known as physiologic races (Stakman, 1917). Oort in the Netherlands (Oort, 1944) and Flor in the USA (Flor, 1942) described the genetic interrelationship between host and pathogen which since then has become known as the gene-for-gene hypothesis. Oort published his findings during World War II in the Dutch language in the journal *Nederlands Tijdschrift over Plantenziekten*, whereas Flor published his results in English in the journal *Phytopathology*. Flor's work became best known and has been cited ever since. He showed that resistance in flax to the fungal pathogen *Melampsora lini*, the causal agent of flax rust, inherits as a dominant monogenic factor (Flor, 1942). For every gene determining resistance in the host, there is a corresponding gene conditioning avirulence in the pathogen. By carrying out several crosses and assessment of the inheritance of avirulence, Flor was able to identify the genotype of the different physiologic races of the rust fungus.

In subsequent years, several plant–pathogen interactions were found to fit patterns complying with the gene-for-gene concept (Crute, 1985) and various models have been proposed to explain the molecular basis of the relationship. The current view is that resistance is generally associated with the so-called hypersensitive response (HR), which is triggered by specific recognition of the pathogen by the host (De Wit, 1986, 1992; Gabriel and Rolfe, 1990; Keen, 1990, 1992; Atkinson, 1993) The HR can be considered as a rapid, programmed death (Greenberg *et al.*, 1994; Dietrich *et al.*, 1994) of a limited number of host cells at the site of penetration and is accompanied by transcriptional activation of various defense genes of the plant. The combination of a rapid collapse of the cells in the vicinity of the invading pathogen and activation of various defense responses in the area surrounding the site of penetration, prevents further spread of the pathogen through the plant tissue. Genetic and biochemical data, obtained from various host–pathogen interactions for which a gene-for-gene relationship has been proposed, support the specific elicitor–receptor model (Gabriel and Rolfe, 1990; Keen, 1990, 1992; De Wit, 1992). The model implies that race-specific elicitors (products of avirulence genes) of the pathogen bind to receptors (putative products of resistance genes) in the host plant. As a result of elicitor binding a signal transduction cascade is activated, which finally results in induction of the HR and activation of host defense genes (Scheel, 1990; Lamb, 1994). It is assumed that resistance gene products are either specific receptors that are directly involved in elicitor perception, or proteins that are associated with these receptors and are involved in one of the first steps of the signal transduction pathway that connects the perception event with the activation of defense responses eventually resulting in resistance.

Cloning and studies on the structure of fungal avirulence genes and the corresponding plant resistance genes will advance our understanding of mechanisms that underly race-specific resistance. Cloned avirulence genes will help us to determine the basis of variation in plant pathogens and their potential to circumvent race-specific resistance, whereas cloned resistance genes will provide insight into the potential of plants to recognize their pathogens. Once both genes and their products are known one can start to study their interactions and the cascade of signaling events that eventually lead to resistance. Cloning of fungal avirulence genes mostly did not parallel cloning the complementary resistance genes. Cloned bacterial avirulence genes have not been instrumental in isolating the corresponding resistance genes. There is, however, the exception of the avirulence gene *Avr9* of *Cladosporium fulvum* which has been helpful in isolating the corresponding *Cf9* resistance gene as will be discussed later (Van den Ackerveken *et al.*, 1992; Jones *et al.*, 1994). Two avirulence genes of *Magnaporthe grisea* have been cloned but nothing is known yet about the complementary resistance genes. The flax rust L^6 has been cloned, but nothing is known yet about its complementary avirulence gene in the flax rust fungus *Melampsora lini*. Cloning of avirulence genes of the plant pathogenic fungi *Rhynchosporium secalis* and *Phytophthora parasitica* and confirmation of their function by transformation is close to completion (W. Knogge, personal communication; B. M. Tyler, personal communication).

This review will mainly focus on fungal avirulence genes and their products, but where appropriate I will also discuss the corresponding resistance genes and resistance gene-dependent defense responses. This field of research develops rapidly and may soon have its payoff in obtaining more durable resistance than the narrow, fragile race-specific resistance which is often overcome by pathogens. This might be achieved by recombining avirulence genes and resistance genes in transgenic plants in such a way that they act against various pathogens rather than against one race of a pathogen only.

II. FUNGAL AVIRULENCE GENES AND THEIR PRODUCTS

To date four fungal avirulence genes and two putative ones have been cloned (Van den Ackerveken *et al.*, 1992; Joosten *et al.*, 1994; Valent and Chumley, 1994; W. Knogge, personal communication; B. M. Tyler, personal communication).

The two fungal avirulence genes of *C. fulvum* have been cloned by a functional biochemical approach. Race-specific elicitor proteins inducing the HR were characterized, and based on the amino acid sequence the encoding genes were cloned. This approach was also followed for the putative avirulence genes from *R. secalis* (Knogge *et al.*, 1994) and *P. parasitica* (Ricci *et al.*, 1989; Kamoun *et al.*, 1993b).

As there was not a convenient HR-based screening system available to isolate race-specific elicitors of *M. grisea* (Valent and Chumley, 1994), and the genome size of this fungus is too large to employ the shotgun approach (which has been used to clone bacterial avirulence genes; Long and Staskawicz, 1993), a map-based cloning approach has been followed to isolate two of its avirulence genes.

The transposon tagging approach, which has been very effective in bacteria to identify and characterize different types of genes (Mills, 1985), can also be employed to isolate fungal avirulence genes, as active transposons have been described in numerous fungi, such as *Fusarium oxysporum* (Daboussi *et al.*, 1992). So far this approach has not yet led to isolation of fungal avirulence genes, however, and it will not be discussed here. A summary of fungal avirulence genes and their products is presented in Table I.

A. AVIRULENCE GENES OF *CLADOSPORIUM FULVUM*

The interaction between the biotrophic fungus *Cladosporium fulvum* and tomato (*Lycopersicon esculentum*), has been studied extensively as a model system for plant–fungus gene-for-gene relationships (De Wit, 1992). In compatible interactions there is no clear response of the host plant tomato upon penetration, allowing the fungus to grow intercellularly in close contact with the mesophyll cells and the vascular tissue (De Wit, 1977; Lazarovits and Higgins, 1976; Van den Ackerveken *et al.*, 1994).

In incompatible interactions, growth of the fungus is arrested immediately after penetration of the leaf. Different near-isogenic lines of tomato cultivar Moneymaker, carrying different single genes for resistance (*Cf2*, *Cf4*, *Cf5* or *Cf9*), give a clear differential response (resistant or susceptible) to the various races of *C. fulvum*. One of the initial defense responses of the host consists of a characteristic HR, which is accompanied by callose deposition and accumulation of phytoalexins (Lazarovits and Higgins, 1976; De Wit, 1977; De Wit and Kodde, 1981). The fast accumulation of various host-encoded pathogenesis-related (PR) proteins in the apoplast is also characteristic for incompatible interactions (De Wit and Van der Meer, 1986; De Wit *et al.*, 1986; Joosten and De Wit, 1989; Kombrink and Somssich, this volume). Biochemical characterization revealed that several of these proteins are 1,3-β-glucanases and chitinases, hydrolytic enzymes potentially able to degrade hyphal walls that contain 1,3-β-glucans and chitin (Joosten and De Wit, 1989). The cDNAs encoding the various basic and acidic 1,3-β-glucanases and chitinases have been cloned (Danhash *et al.*, 1993; Van Kan *et al.*, 1992) and the corresponding genes expressed (Danhash *et al.*, 1993; Van Kan *et al.*, 1992;

TABLE I
Cloned fungal avirulence genes and their products

Avirulence gene	Precursor/mature[a] protein	Activity	Homology	Reference
Avr9	63 aa/28 aa	HR induction in Cf9 genotypes	Promoter *nit-3* gene	Van den Ackerveken *et al.* (1994)
Avr4	135 aa/106 aa	HR induction in Cf4 genotypes	None	Joosten *et al.* (1994)
PWL2	145 aa/?[b]	Not known	None	Valent and Chumley (1994)
AVR2-YAMO	223 aa/?	Not known	Neutral Zn-protease?	Valent and Chumley (1994)
avrRrs1	82 aa/60 aa	Induction of PRHv-1 in barley Atlas 46	None	Knogge *et al.* (1994)
parA1	118 aa/98 aa	Necrosis induction in tobacco, radish and turnip	None	Kamoun *et al.* (1993b)

[a] All avirulence gene products are secreted proteins (see text for details).
[b] Mature protein not known yet.

Wubben *et al.*, 1994, 1995); localization of the transcripts and the encoded proteins has been studied (Wubben *et al.*, 1992, 1994, 1995). Although the early accumulation of the hydrolytic enzymes in the incompatible interaction coincides with the expression of HR and arrest of fungal growth, it is not clear whether the induced PR proteins indeed play a decisive role in resistance of tomato against *C. fulvum* (Joosten *et al.*, 1995).

The confinement of fungal growth to the apoplast allowed the isolation of race-specific elicitors of apoplastic fluid (AF) from compatible interactions which eventually led to the cloning of the first fungal avirulence gene. AF which is obtained by *in vacuo* infiltration of tomato leaves with water, followed by low speed centrifugation (De Wit and Spikman, 1982), contains proteins that are constitutively produced by the host and the pathogen, as well as proteins which are synthesized as a result of the interaction between host and pathogen (De Wit *et al.*, 1986, 1988). In the search for proteinaceous race-specific elicitors that induce *Cf*-specific defense responses, AF isolated from compatible interactions involving races carrying different avirulence genes was analysed for the presence of these compounds.

1. Race-specific elicitors and cloning of their encoding avirulence genes
The first evidence for the presence of race-specific elicitors in AF isolated from compatible interactions between tomato and *C. fulvum* was obtained by De Wit and Spikman (1982) and was confirmed later (De Wit *et al.*, 1984, 1985). Circumstantial evidence was obtained for the occurrence of four proteinaceous elicitors in AF, the putative products of avirulence genes *Avr2*, *Avr4*, *Avr5* and *Avr9*. The AVR4 and AVR9 peptide elicitors have been purified and their amino acid sequences have been determined (Scholtens-Toma and De Wit, 1988; Joosten *et al.*, 1994). The AVR9 peptide consists of 28 amino acids, including six cysteine residues. Based on the amino acid sequence of the elicitor, degenerate oligonucleotide probes were designed and used to screen a cDNA library prepared from *C. fulvum*-infected leaf tissue (Van Kan *et al.*, 1991). The primary structure of mRNA encoding the AVR9 elicitor revealed that the primary gene product is a pre-pro-protein of 63 amino acids, which contains the sequence of the mature elicitor at the carboxy-terminus. The pre-pro-protein has a signal peptide of 23 amino acids, which is cleaved off upon extracellular targeting. The resulting extracellular peptide of 40 amino acids is N-terminally processed by fungal proteases into peptides of 32, 33, or 34 amino acids, respectively, which accumulate in culture filtrates of transformants of *C. fulvum* that constitutively produce the AVR9 elicitor (Van den Ackerveken *et al.*, 1993). When the purified peptide of 34 amino acids is incubated with AF isolated from uninfected tomato leaflets, it is further processed into the mature elicitor of 28 amino acids, indicating that plant proteases are necessary for the final processing (Van den Ackerveken *et al.*, 1993). However, it cannot be excluded that plant proteases are able to process the pro-protein of 40 amino acids directly into the mature elicitor peptide of

28 amino acids. It is also not clear whether the proteolytic processing steps that take place after extracellular targeting of the peptide are essential for necrosis-inducing activity of the AVR9 elicitor.

The race-specific AVR4 elicitor, which specifically-induces necrosis in tomato genotypes that carry resistance gene *Cf4*, has been isolated and purified in a similar way as the AVR9 peptide elicitor (Joosten *et al.*, 1994). Analysis of *Avr*4 cDNA revealed that, similar to the AVR9 peptide, the AVR4 protein is encoded as a pre-pro-protein, with a putative amino-terminal signal peptide of 18 amino acids. Analogous to AVR9, an additional stretch of 11 amino acids at the N-terminus is most likely cleaved off by plant and/or fungal proteases after extracellular targeting. The mature AVR4 protein of 106 amino acids contains eight cysteine residues and shares no significant homology to AVR9 or other proteins present in various databases.

In order to prove that the cloned *Avr*4 and *Avr*9 genes indeed determined avirulence, virulent strains were transformed with the respective genes and their phenotypes on the Cf4 and Cf9 genotypes were determined. Both genes indeed proved to be genuine avirulence genes as they changed the phenotype of these stains into avirulence (Van den Ackerveken *et al.*, 1992; Joosten *et al.*, 1994). Disruption of the *Avr*9 coding sequence in wild-type races of *C. fulvum*, rendered these races virulent on tomato genotypes that contain resistance gene *Cf9* (Marmeisse *et al.*, 1993).

These results confirm that the cloned *Avr*4 and *Avr*9 genes are the only factors determining avirulence of *C. fulvum* on tomato varieties that carry the *Cf4* and *Cf9* resistance genes, respectively, and that they comply with the definitions of avirulence genes.

2. Avr-*Alleles in wild-type races of* C. fulvum *virulent on tomato genotypes* Cf4 *and* Cf9

Southern analysis, with the *Avr*4 cDNA as a probe, did not reveal any difference between races of *C. fulvum* avirulent or virulent on tomato genotype Cf4. All races contained a homologous, single copy gene, not displaying any restriction fragment length polymorphism with the restriction enzymes that were used (Joosten *et al.*, 1994). Although none of the virulent races produces biologically active AVR4 elicitor, RNA isolated from those races grown on a susceptible tomato genotype contains transcripts that hybridize to an *Avr*4 cDNA probe, proving that those races containing "virulent" alleles of *Avr*4 are transcribed. Sequencing of the coding region revealed only single base pair mutations in the ORF encoding the mature AVR4 protein. These mutations resulted in single amino acid changes in the encoded peptide. The amino acid changes involve mainly changes from cysteine to tyrosine, but also changes from a tyrosine to histidine and from a threonine to isoleucine have been found. In one race virulent on tomato genotype Cf4, a frame shift mutation was observed leaving only 13 amino acids at the N-terminus of the wild-type ORF of the *Avr*4 gene intact (Joosten *et al.*, 1994; M. H. A. J. Joosten *et al.*, unpublished). The consequence of the single codon changes for biological activity of the AVR4 protein suggests that altering secondary and tertiary structure of the AVR4 elicitor peptide might inhibit or prevent binding of the peptide to a complementary receptor in the resistant cultivar. Alternatively, the structural change in the AVR4 protein might cause the peptide to become more sensitive to proteolytic degradation which would prevent it from reaching a receptor. A third possibility could be that the mutated peptide is not secreted by the virulent races. Preliminary studies by M. H. A. J. Joosten *et al.* (unpublished) indicate that proteins produced by virulent strains are unstable, suggesting that virulent strains behave phenotypically as low or non-producers of potentially active elicitors. It is also possible that the elicitors are not secreted by these races.

Southern analysis of many races of *C. fulvum* virulent on tomato genotype Cf9 revealed that in these races the *Avr9* gene was always absent. Thus absence of the *Avr9* gene and virulence on Cf9 genotypes was always found to be correlated (Van Kan *et al.*, 1991).

3.   Regulation and possible intrinsic functions of avirulence genes Avr4 and Avr9

Northern analysis indicated that expression of the *Avr4* and *Avr9* genes of *C. fulvum* is specifically induced *in planta*. Accumulation of mRNAs encoding the race-specific elicitors correlates with the increase in fungal biomass in the tomato leaves during pathogenesis in compatible interactions (Joosten *et al.*, 1994; Van Kan *et al.*, 1991).

Analysis of the promoter of the *Avr4* gene for the presence of specific motifs did not reveal any significant homology to sequences known for binding of regulatory proteins. Analysis of transformants of *C. fulvum* carrying *Avr4* promoter-GUS fusions *in planta*, should reveal at which particular stage(s) during pathogenesis expression of the gene is induced (M. H. A. J. Joosten *et al.*, unpublished).

In contrast to the *Avr9* gene, which is absent in races of *C. fulvum* virulent on tomato genotype Cf9, the presence of mutated *Avr4* alleles in several races of *C. fulvum*, might suggest an essential role for its product in pathogenicity. However, the isolate with a frame-shift mutation in the ORF encoding an AVR4 homolog of only 13 amino acids did not show impaired pathogenicity, indicating that the AVR4 protein is dispensable (M. J. Joosten *et al.*, unpublished results).

When *C. fulvum* is grown in liquid medium, expression of the *Avr9* gene could be induced under limiting concentrations of nitrogen (present as nitrate, ammonium, glutamate or glutamine; Van den Ackerveken *et al.*, 1994). Analysis of the *Avr9* promoter sequence revealed several potential regulatory elements in two direct repeats. The promoter region contains six copies of the hexanucleotide sequence TAGATA (Van den Ackerveken *et al.*, 1994), which was identified as the recognition site of the NIT2 and AreA proteins, transcription factors that positively regulate gene expression under nitrogen-limiting conditions in *Neurospora crassa* and *Aspergillus nidulans*, respectively (Fu and Marzluf, 1990; Caddick, 1992). Possibly, expression of *Avr9* is regulated by a *C. fulvum*-homolog of the NIT2 protein in a similar way. Deleting several of these TAGATA boxes abolished the induction of the *Avr9* gene under conditions of low nitrogen concentration *in vitro* (P. J. G. M. De Wit *et al.*, unpublished).

Studies on the expression of the *Avr9* gene *in planta*, using transformants of *C. fulvum* carrying *Avr9* promoter–GUS fusions, showed activation of the promoter in hyphae after the fungus passed the stomatal guard cells, whereas expression was particularly high in mycelium growing in the vicinity of the vascular tissue (Van den Ackerveken *et al.*, 1994).

The observations that transformants of *C. fulvum* in which the *Avr9* gene was disrupted, did not show impaired pathogenicity in a monocyclic infection assay, and that in nature avoidance of *Cf9*-specific resistance is achieved by complete deletion of the *Avr9* gene, suggest that the *Avr9* gene is dispensable for growth and pathogenicity of *C. fulvum*. The *Cf9* resistance gene, however, which has been present in tomato breeding lines since 1979, still renders complete protection to *C. fulvum* in commercial tomato crops, indicating that loss of the *Avr9* genes, and maybe of linked genes, interferes with competitive abilities of *C. fulvum* in nature. Perhaps the *Avr9* gene plays a role in the epidemiology of the disease. Additional research, using near-isogenic races of *C. fulvum* that are *Avr9⁻* or *Avr9⁺* in polycyclic infection experiments, should reveal whether this explanation is feasible.

4. Relation between structure and biological function of AVR4 and AVR9 peptides

Both avirulence genes *Avr4* and *Avr9* encode relatively small globular peptide elicitors which contain eight and six cysteine residues, respectively. The structure of the AVR9 peptide has been extensively studied by ¹H-NMR (J. J. M. Vervoort *et al.*, unpublished). All cysteines form disulfide bridges, which are needed for elicitor activity as the fully reduced molecule is no longer active. The protein contains three anti-parallel β-sheets and two loops of two and 10 amino acids. Interaction between the amino acids or the three antiparallel β-sheets is strong; they form a barrel-like structure the interior of which is extremely hydrophobic and strongly inhibits the free exchange of protons. In order to get more insight into the region of the molecule which interacts with a putative receptor, *in vitro* mutagenesis is being performed on the *Avr9* gene to obtain AVR9 elicitor molecules with altered biological activities. Two approaches are being used: first, mutant constructs of the 28 amino acid peptide encoding the *Avr9* gene differing in only one amino acid containing the PR1a signal peptide were made and cloned behind the coat protein promoter potato virus X (PVX) as described by Hammond-Kosack *et al.* (1994). Transcripts were made *in vitro* and inoculated onto tobacco or tomato genotypes with or without the resistance gene *Cf9*. When an active AVR9 peptide is produced the tomato Cf9 genotype becomes heavily necrotic due to a constitutive HR and dies within a week. Active or inactive AVR9-like peptides would show little or no necrosis, respectively, whereas highly active AVR9-like peptides would kill *Cf9* genotypes of tomato more quickly than the wild-type AVR9 elicitor. This is indeed what has been found by studying various mutant AVR9-like peptides (R. Vogelsang *et al.*, unpublished). In a second approach, a few mutagenized *Avr9* genes will be expressed constitutively in *C. fulvum* in order to obtain sufficient amounts of mutant AVR9 elicitors for ^{1}H-NMR studies. This will provide information on the effect of the mutations on the three-dimensional structure of the peptide which is vital for elicitor activity.

The availability of mutant forms of AVR9 peptides is also crucial for physical characterization of the AVR9 receptor in tomato and possibly other plants (M. Kooman-Gersmann *et al.*, unpublished). In studies on mutant versions of the *Avr9* gene, the codon preference of tomato was used (R. Vogelsang *et al.*, unpublished). In recent studies on the AVR4 peptide it was found that the non-optimized, authentic wild-type fungal *Avr4* coding sequence, inserted in the PVX vector, generated active AVR4 elicitor in tomato plants. Cf4 plants were killed indicating that the plant can produce the 10 kDa fungal AVR4 elicitor (R. Vogelsang *et al.*, unpublished). The latter result is significant as no large synthetic constructs of the *Avr4* gene are required for structure-function relation studies. *In vitro* mutagenesis can now be carried out on the wild-type *Avr4* construct by using the PVX vector.

As indicated above, structure–function relationship studies on AVR proteins will help us to unravel physical aspects of ligand–receptor interactions. This type of study is relevant since the resistance gene *Cf9* corresponding to the avirulence gene *Avr9* has been cloned recently (discussed later).

5. Stable integration and expression of avirulence genes Avr4 *and* Avr9 *in tobacco and tomato*

Expressing avirulence genes under different levels of control creates various ways to study the effects of their encoded race-specific elicitors in plants carrying the corresponding genes for resistance. Physiological and biochemical responses elicited by constitutively induced HR or pathogen-induced HR can be studied. The constitutive expression of the *Avr9* gene in plants has been instrumental for cloning the complementary resistance gene *Cf9* (Jones *et al.*, 1994). In the previous section it was discussed that plants can produce biologically active AVR4 and AVR9 elicitor. As the AVR9 elicitor is secreted into the apoplast by *C. fulvum* during its interaction with tomato, and plants encounter this elicitor via the apoplast, it is important to show that transgenic plants should also secrete elicitors into the apoplast. To this end, a synthetic *Avr9* gene was constructed initially, which was preceded by the signal peptide

(SP) of the tobacco pathogenesis related proteins PRS or PR1a (G. Honée *et al.*, unpublished; Hammond-Kosack *et al.*, 1994). Constructs were made which encode a 63, 40, or 28 amino acid peptide elicitor protein representing the presumed pre-pro-protein, the pro-protein or the mature elicitor protein, respectively (G. Honde *et al.*, unpublished). The ORFs were placed under the control of either the constitutive 35S promoter or pathogen-inducible promoters. The *35S:SP:Avr9* constructs were made in a plant transformation vector containing the neomycin phosphotransferase gene as a selection marker. Transformation was achieved with *Agrobacterium tumefaciens* by use of the binary vector system. In this way transgenic tobacco and tomato plants were generated which contained different stably integrated constructs of the *Avr9* gene and produce biologically active elicitor molecules (G. Honée *et al.*, unpublished; Hammond-Kosack *et al.*, 1994). All transgenic plants produced the mature 28 amino acid AVR9 elicitor, indicating that the plants are able to perform the whole processing of the 63 amino acid pre-pro-protein, while transgenic *C. fulvum* grown *in vitro* and constitutively expressing the 63 amino acid pre-pro-protein encoding *Avr9* gene is only able to produce a mixture of 32, 33, and 34 amino acid peptides (Van den Ackerveken *et al.*, 1993).

Crossing transgenic AVR9 producing Cf9$^-$ genotypes with Cf9$^+$ genotypes generated viable tomato seeds (Hammond-Kosack *et al.*, 1994). Apparently the presence of the *Cf9* resistance gene and the active *Avr9* gene in one plant does not cause deleterious effects on flower deployment, fertilization, embryo and fruit development and seed maturation (Hammond-Kosack *et al.*, 1994). However, when seeds from such F1 transgenic plants were germinated, clear necrosis became visible on cotyledons and young developing leaves of the seedlings. Use of a comparable *35S:Gus* construct revealed that these would have been expression of *Avr9* in all tissues except the conducting tissue of the style. As HR only occurs in leaves, this indicates that either: (i) the resistance gene *Cf9* is not expressed in tissues which do not show spontaneous necrosis; (ii) the tissues are not competent to respond to the elicitor; or (iii) there are factors present in these tissues, which prevent the initiation of necrosis (Hammond-Kosack *et al.*, 1994). However, in future the expression of the *Cf9* gene can be followed in detail with a *Cf9* probe on northern blots containing RNAs from different plant tissues as the resistance gene has been cloned recently (Jones *et al.*, 1994). A fourth possibility could be that for generation of necrosis two components are required; one could be the AVR9 receptor which does not have to be the product of the resistance gene itself and which is possibly not expressed in all tissues, whereas the other could be the resistance gene product itself, which might encode an essential component in the signal transduction pathway required for the eventual development of HR. It still has to be determined in detail whether tissues not visibly undergoing deleterious effects indeed show no other defense-related responses. The visible deleterious effects in cotyledons and leaves eventually leading to

Avr9-conditioned HR are developmentally regulated and expressed at the time stomata become fully functional. Functional stomata respond by supra-optimal opening as a result of AVR9 elicitor action (Hammond-Kosack *et al.*, 1994), which leads to a quick HR.

B. AVIRULENCE GENES OF *MAGNAPORTHE GRISEA*

The fungal pathogen *Magnaporthe grisea* causes blast of rice and different grass species. It is renowned for its physiological specialization on rice (Silué *et al.*, 1992a). New races rapidly appear when new commercial blast-resistant cultivars are introduced. The rice blast researchers consider the rice blast system to represent a gene-for-gene relationship. It is thus anticipated that the rice pathogen is capable of infecting a particular rice cultivar only if the patho-gen lacks all the *Avr* genes that correspond to the cultivars resistance genes, or when the particular cultivar lacks all resistance genes corresponding to the fungal avirulence genes. Evidence that the rice blast system has a gene-for-gene basis came from reports on Mendelian segregation of single genes that control cultivar specificity in the pathogen, and corresponding dominant resistance genes in the host (Silué *et al.*, 1992a, b).

The molecular basis of host specificity in the rice blast disease has been studied intensively in recent years (Valent and Chumley, 1991, 1994; Silué *et al.*, 1992a, b). Analysis of mutation events in two unstable avirulence genes, *PWL2* and *AVR2-YAMO* facilitated the map-based cloning of these two genes.

1. Species-specific avirulence gene PWL2

Some rice isolates of *M. grisea* that are non-pathogenic on weeping lovegrass, undergo frequent spontaneous mutation to pathogenicity towards this host. Similarly, some rice isolates that are avirulent on rice cultivars, such as Yashiro-mochi, mutate frequently to virulence on this cultivar. The *PWL2* gene, originating from such a rice isolate and causing non-pathogenicity on weeping lovegrass, has been cloned by chromosome walking, starting with a physical marker (Valent and Chumley, 1994; J. A. Sweigard *et al.*, unpublished). However, walking appeared to be difficult due to presence of repetitive DNA sequences. Attempts to isolate single copy sequences to use as hybridization probes for identifying overlapping cosmid clones were unsuc-cessful. Potential single copy probes were proven to be low copy repeat sequences present at four to six unlinked locations in the *M. grisea* genome.

Sometimes DNA sequences linked to *PWL2* in an RFLP mapping popula-tion were missing in mutants which had become pathogenic on weeping lovegrass. Cosmids suspected to contain *PWL2* were used to transform strains pathogenic on weeping lovegrass and appeared to be able to generate transformants non-pathogenic on weeping lovegrass. This was the proof for having cloned a functional species-specific non-pathogenicity or avirulence gene. Further analysis of the *PWL2* gene revealed that it encodes a glycine-rich protein of 145 amino acids. The putative signal peptide suggests that the protein is secreted. The protein shows no homology to proteins present in databases. The instability at the *PWL2* locus is mainly due to its chromosomal location, which causes different types of deletions at the tip of the chromosome as large as 30 kb. One mutant appeared to be different as it carried only a single base pair change in the ORF. The protein encoded by *PWL2*, produced in sufficient amounts *in vitro*, does not induce an HR-response in weeping lovegrass (B. Valent, personal communication), indicating that it does not code for an elicitor molecule such as the *Avr4* and *Avr9* genes of *C. fulvum*.

2. Cultivar-specific avirulence gene AVR2-YAMO

The *AVR2-YAMO* gene, when present in isolates of *M. grisea*, prevents infection of the rice cultivar Yashiro-mochi. By mutant analysis it was discovered that this gene resides near the tip of a chromosome (Valent and Chumley, 1994; M. J. Orbach *et al.*, unpublished). Spontaneous mutants virulent on cultivar Yashiro-mochi showed structural changes in telomeric restriction fragments that mapped with AVR2-YAMO. In eight of eleven independently-isolated spontaneous mutants, loss of expression of AVR2-YAMO correlated with the disappearence of the linked telomeric DNA fragment. *AVR2-YAMO* appears to reside within 1–2 kb of the chromosomal tip. Indeed, transfer of a telomeric 6.5 kb *Bgl* II DNA fragment into a strain virulent on Yashiro-mochi led to transformants that were unable to infect this cultivar and thus proved that this DNA fragment contained the *AVR2-YAMO* avirulence gene.

The *AVR2-YAMO* gene encodes a protein of 223 amino acids with no significant homology to other known proteins in databases. Unlike the relatively uniform deletion events leading to virulence and occurring at the unstable *PWL2* locus, many different genetic events can result in inactivation of the *AVR2-YAMO* gene. These include deletions varying from 100 bp up to 12.5 kb, insertions and point mutations (Valent and Chumley, 1994; M. J. Orbach *et al.*, unpublished).

3. Homologs of PWL2 and AVR2-YAMO in different plant pathogens

A search for homologous genes led to the discovery of three other *PWL2*-homologs, *PWL1*, *PWL3* and *PWL4* (S. Kang *et al.*, unpublished). The *PWL1* and *PWL3* genes were isolated from a goosegrass isolate based on screening with the homologous *PWL2* probe, *PWL4* was isolated from a weeping lovegrass pathogen. *PWL1* and *PWL4* encode functional proteins but *PWL3* does not. Functional homologs of the *AVR2-YAMO* gene are present in *M. grisea* strains pathogenic on grasses such as *Digitaria* and *Pennisetum* (Valent and Chumley, 1994; S. Kang *et al.*, unpublished). These data show

that the occurrence of species- and cultivar-specific avirulence genes is not
limited to isolates from rice.

4. Biological functions of avirulence genes of M. grisea

Clearly, avirulence genes in the pathogen did not evolve to prevent the pathogen from infecting other host species or various genotypes of host species. It
is likely that host plants have become able to detect molecules of pathogens
with different biological functions. The importance of such a function for the
pathogen determines whether an virulence gene can be lost without a fitness
penalty. It is expected that molecules which act as avirulence genes, and at
the same time are crucial for the pathogen, cannot easily be lost by the
pathogen. The proteins encoded by *PWL2* and *AVR2-YAMO* have no significant homology to amino acid sequences of proteins with known functions.
However, a short stretch of amino acids in the sequence of *AVR2-YAMO*
shares homology with the active center of a neutral Zn-protease. In virulent
strains in *AVR2-YAMO* a few point mutations were found in this putative
active center suggesting that for avirulence the active center should be intact
(B. Valent, personal communication). Whether the *AVR2-YAMO* gene product indeed possesses protease activity still remains to be determined. However, both genes can be lost and can provide *M. grisea* the possibility to enlarge
its host range. No data are available on the effects of deletions of *PWL2* and
AVR2-YAMO on the fitness of the pathogen, apart from becoming virulent
on the appropriate cultivars.

C. THE AVIRULENCE GENE *avrRs1* OF *RHYNCHOSPORIUM SECALIS*

Rhynchosporium secalis causes barley leaf scald. Resistance toward this
pathogen is governed by various dominant resistance genes (Dyck and Challer,
1961; Habgood and Hayes, 1971). Resistance gene *Rrs1* is regarded as part
of a complex of closely linked genes including *Rrs3* and *Rrs4*. *Rrs1* is available
in cultivar Atlas 46, whereas the near-isogenic line Atlas does not contain this
resistance gene. Atlas 46 (*Rrs1*) provides resistance toward fungal race
US238.1, whereas race AU2 grows abundantly on this cultivar. Resistance is
not caused by a hypersensitive response. Penetration proceeds in a very similar
way in susceptible and resistant cultivars leading to an early collapse of a few
epidermal cells. In the susceptible cultivar fungal growth continues, giving rise
to large necrotic sporulating lesions, whereas in the resistant cultivars fungal
growth stops (Lehnackers and Knogge, 1990). The interaction fits a gene-for-
gene model, where resistance is semi-dominant (Hahn *et al.*, 1993). The fungus
produces a number of necrosis-inducing peptides NIP1, NIP2, and NIP3
(Wevelsiep *et al.*, 1991), of which NIP1 has the properties of a race-specific
elicitor as will be discussed below.

1. Isolation of the NIP1 elicitor and cloning of its encoding gene
A small family of necrosis-inducing peptides (NIPs) was identified in culture filtrates of *R. secalis*, as well as in infected susceptible plants (Wevelsiep *et al.*, 1991). One of these peptides, NIP1, was detected only in culture filtrates of the fungal race US238.1, which is avirulent on cultivars carrying resistance gene *Rrs1*. This necrosis-inducing peptide was found to specifically elicit the accumulation of mRNA encoding the barley PR protein PRHv-1 in an *Rrs1*-containing cultivar in a similar way to the avirulent race US238.1 (Hahn *et al.*, 1993). This property made NIP1 a candidate for a race-specific elicitor being the putative product of fungal avirulence gene *avrRrs1*, interacting with the corresponding resistance gene *Rrs1*. NIP1 is a protein of 60 amino acids, with 10 cysteine residues. Based on the amino acid sequence of the NIP1 elicitor, oligomers were designed that enabled the isolation of a cDNA clone encoding the NIP1 elicitor (Knogge *et al.*, 1994).

2. Circumstantial evidence for avrRrs1 *being an avirulence gene*
The research group of Knogge is presently transforming races virulent on the barley cultivar Atlas 46, containing resistance gene *Rrs1*, with the *avrRrs1* gene. If *avrRrs1* is a genuine avirulence gene, those races should become avirulent on Atlas 46. The outcome of the experiment is not yet known. Additional confirmation that *avrRrs1* is a genuine avirulence gene would be provided if disruption of the NIP1-encoding gene in race US238.1 would render this race virulent on Atlas 46. In the meantime Knogge and co-workers demonstrated that co-inoculation of virulent race AU2 with the NIP1 elicitor protein on Atlas 46 resulted in an incompatible interaction, indicating that NIP1 induces defense responses which prevent further growth of the virulent race (W. Knogge, personal communication).

3. Occurrence and expression of the avrRrs1 *gene in different isolates of R. secalis*
When genomic DNA of 12 isolates of *R. secalis* was analysed for presence of the *avrRrs1* gene, it was shown that all isolates avirulent on cultivars carrying *Rrs1* contained an *avrRrs1* homolog, and nearly all races virulent on *Rrs1* cultivars lacked the homolog. One exception was the occurrence of an *avrRrs1* allele in the virulent race AU2. Therefore, the coding sequences of *avrRrs1* homologs from eight fungal races were compared. All *avrRrs1*-homologous genes encode mature proteins of 60 amino acids. Six of the genes differ in the same three nucleotide positions from the *avrRrs1* gene of race US238.1. The resulting three amino acid exchanges do not appear to influence elicitor activity of the gene products. However, the *avrRrs1* product of race AU2 is characterized by a fourth amino acid exchange, leading to loss of elicitor activity. An additional race, AU3, whose interaction with *Rrs1* plants is not governed by the *avrRrs1* gene as it is also avirulent on cultivar Atlas (lacking the *Rrs1* gene), expresses an NIP1 protein with a fourth amino acid exchange

at a position different from race AU2. This protein also lacks elicitor activity. These results indicate that single amino acid exchanges in the NIP1 protein are sufficient to change the phenotype from avirulent to virulent. These results are similar to those observed for the avirulence gene *Avr4* of *C. fulvum* (Joosten *et al.*, 1994; M. H. A. J. Joosten *et al.*, unpublished).

4. Possible biological functions of NIP1

NIP1 is produced during the infection process and is therefore expected to have a biological function during pathogenesis. The phytotoxicity of NIP1 as well as of NIP2 and NIP3 might be an important feature of pathogenicity of the fungus during growth in plants lacking the *Rrs1* gene. NIP1 and NIP3 are reported to stimulate H^+-ATPase activity in barley (Wevelsiep *et al.*, 1993). On susceptible cultivars, races of *R. secalis* that lack the *avrRrs1* gene (encoding NIP1) develop symptoms more slowly than races which contain the *avrRrs1* gene. This indicates that the *avrRrs1* gene is not absolutely required for pathogenicity but is of importance for symptom development. This suggests that the intrinsic function of the *avrRrs1* gene could be related to pathogenicity.

D. THE PUTATIVE AVIRULENCE GENE *parA1* OF *PHYTOPHTHORA PARASITICA*

Many *Phytophthora* species produce a family of small extracellular proteins of ca. 10 kDa named elicitins, which induce necrosis and other defense responses particularly on tobacco and some radish and turnip cultivars (Ricci *et al.*, 1989, 1992; Pernollet *et al.*, 1993; Kamoun *et al.*, 1993a, b). The response to elicitins induces resistance against subsequent infection by silica on tobacco and by the bacterial pathogen *Xanthomonas campestris* pv. *amoraciae* on radish. Elicitins are named after the *Phytophthora* species by which they are produced, e.g. *P. parasiticeia* and cryptogein are produced by *P. parasitica* and *P. cryptogea*, respectively. Based on structure and biological activity, elicitins can be classified into two groups: the α-elicitins which are acidic proteins (e.g. parasiticein) and the β-elicitins which are basic hydrophilic proteins (e.g. cryptogein; Nespoulos *et al.*, 1992). Some isolates of *P. parasitica* that are virulent on tobacco do not produce parasiticein, whereas many isolates of *P. parasitica* that are non-pathogenic on tobacco do produce parasiticein (Ricci *et al.*, 1989, 1992). As elicitins have the property to induce necrosis mainly on tobacco, they could thus be considered as species-specific elicitors. They may block or slow down infection by triggering a defense response in the host (Kamoun *et al.*, 1993a; Ricci *et al.*, 1989, 1992). A cross between a pathogenic non-parasiticein producer and a non-pathogenic parasiticein producer gave surprising results. Only four of 21 progeny produced parasiticein, suggesting a complex genetic control of parasiticein production (Kamoun *et al.*, 1994). The parasiticein producers were all non-pathogenic, but the non-producers varied from highly pathogenic to low pathogenic. The non-producers varied strongly in growth rate *in vitro*, whereas all producers showed a high growth rate *in vitro*, suggesting that parasiticein production might be a fitness factor. These results suggest that parasiticein production confers only part of the non-pathogenicity on tobacco or that pathogenic parasiticein producers contain suppressors of parasiticein action (Kamoun *et al.*, 1994). The genetic study suggests, albeit not strongly, that parasiticein can be considered as an avirulence factor in the *P. parasitica*–tobacco interaction (Ricci *et al.*, 1992; Kamoun *et al.*, 1993a).

1. Cloning of parA1, the gene encoding parasiticein

The *parA1* gene has been cloned from *P. parasitica* by using oligonucleotides that were synthesized based on the amino acid sequence of the purified parasiticein (Kamoun *et al.*, 1993a, b). The *parA1* gene encodes a pre-protein of 118 amino acids, including a signal peptide of 20 amino acids. The mature parasiticein contains 98 amino acids of which six are cysteine residues. In *P. parasitica* the *parA1* gene is a member of a complex multigene family, which comprises at least two additional homologous genes, *parA2* and *parA3*. Whether gene products of *parA2* and *parA3* are involved in the interaction with the plant remains unknown (B. M. Tyler, personal communication). However, when expressed in *E. coli* the elicitin domains of *parA2* and *parA3* yielded active HR elicitors. The *parA1* gene is present in parasiticein producers and non-producers. Possibly, the absence of elicitin production in virulent strains is due to *cis*-acting mutations or mutations in *trans*-acting regulatory proteins. Genetic crosses between parasiticein producers and non-producers did not give a clear answer concerning the role of parasiticein as an avirulence factor due to different genetic backgrounds in the parent lines (Kamoun *et al.*, 1994). Therefore the proof that the *parA1* gene is a genuine species-specific avirulence gene can only be demonstrated by transforming a virulent, non-producing isolate of *P. parasitica* with the *parA1* gene. If transformants indeed become non-pathogenic on tobacco, the *paral* can be considered a genuine avirulence gene. However, transformation of *P. parasitica* appears to be extremely difficult (B. M. Tyler, personal communication).

2. Possible biological functions of elicitins

Production of elicitins is almost ubiquitous among *Phytophthora* species. Of 85 *Phytophthora* isolates representing 14 species, 87% produced elicitins (Kamoun *et al.*, 1994) which suggests that these proteins could play an important role in the genus *Phytophthora*. As the elicitins are produced *in planta*, during compatible interactions they might represent a pathogenicity factor. However, the elicitins could also play a role during the saprophytic phase of *Phytophthora* species. Parasiticein non-producers of *P. parasitica* showed a slower growth rate *in vitro* than parasiticein producers (Kamoun *et al.*, 1994). This suggests that parasiticein in addition to being an avirulence factor, may have functions during parasitic and saprophytic phases (Kamoun *et al.*, 1994).

The elicitin capsicein, produced by *P. capsici*, has some structural homology to phospholipase A_2 (Huet *et al.*, 1994) and phospholipase tests of capsicein proved to be positive. The biological relevance of this activity for *P. capsici* is yet unknown.

III.   DEFENSE RESPONSES INDUCED BY RACE-SPECIFIC ELICITORS

The term elicitor was initially coined for compounds of pathogen origin which are able to induce the accumulation of phytoalexins, one of the defense responses studied intensively in the 1970s and early 1980s (Keen, 1975, 1982). However, currently, the term elicitor is used for all compounds able to induce any defense response including the HR, accumulation of phytoalexins, lignification, cell wall thickening, callose deposition or induction of various enzymes involved in defense responses (Dixon *et al.*, 1994; Kombrink and Somssich, this volume).

The HR is one of the most frequently occurring defense responses induced by specific fungal elicitors. The HR, which generally refers to rapid cell death, is common to various plant species that show active resistance to viral, bacterial, or fungal pathogens (Kiraly, 1980). No experimental studies have definitely identified the cause of the HR. Genetic studies based on mutational analysis in *Arabidopsis thaliana* provided evidence that the HR might be part of a programmed cell death response that is generated by plants after attack by different pathogens (Dietrich *et al.*, 1994; Greenberg *et al.*, 1994). Some mutants showed spontaneous lesions that resembled the HR and most of the defense responses associated with it, as observed during infection of plants by an avirulent race of a pathogen. Therefore, it is assumed that the HR has a similar genetic and physiological basis in different plant–pathogen interactions (Atkinson, 1993). Of the elicitors described above, AVR4 and AVR9 induce the HR, NIP1 and parasiticein induce HR-like necrosis, whereas PWL2 and AVR2-YAMO do not induce a visible HR response. Defense gene activation requires *de novo* mRNA and protein synthesis, whereas very early responses, such as the release of active oxygen species (AOS) which occurs within minutes after treatment with race-specific elicitors, do not seem to require *de novo* mRNA and protein synthesis (Legendre *et al.*, 1992). In comparison with the extensive biochemical studies on plants and plant cell suspensions treated with non-specific elicitors (Atkinson, 1993; Nürnberger *et al.*, 1994; Ebel and Cosio, 1994), data on early biochemical responses induced by fungal race-specific elicitors such as AVR4 and AVR9 and the putative specific elicitors NIP1 and parasiticein are still very limited.

A. RESPONSES INDUCED BY ELICITORS OF C. FULVUM

Peever and Higgins (1989) injected a crude preparation of race-specific elicitor AVR9 of *C. fulvum* into leaflets of the resistant tomato cultivar Sonatine heterozygous for *Cf9*. A specific, elicitor-dependent, increase in electrolyte leakage was observed. This leakage coincided with an increase in lipoxygenases at 6 h, which peaked at 24 h after injection and returned to basal levels by 48 h after treatment, while the first significant lipid peroxidation occurred at 12 h. Similar results were obtained by Hammond-Kosack and Jones (1995), who used cotyledons of 14-day-old tomato seedlings. Upon injection of apoplastic fluid (AF) containing the AVR9 elicitor into cotyledons of Cf9 plants, increased levels of total and oxidized glutathione were observed within one 1–2 h, followed by an oxidative burst by 3 h. By about 9–10 h, ethylene production and loss of membrane integrity became evident. From 12 h onwards a significant increase in free salicylic acid occurred which peaked at 24 h at a concentration of 18 μg/g fresh weight. Similar responses were found in leaves of Cf2 genotypes treated with AF containing AVR2 elicitor, with only slightly different timing. A significant difference between the two cultivars was the supra-optimal opening of the stomata in Cf9 plants at 4 h after treatment which was not observed in treated Cf2 plants.

Vera-Estrella *et al.* (1992, 1994a, b) extended studies on specific biochemical responses to cell suspension cultures. Cell suspension cultures, initiated from callus obtained from tomato genotypes Cf4 and Cf5, retained the specificity of the intact plants from which they originated. Within 10 minutes after treatment with AF containing matching elicitor activity, a marked increase in the extracellular production of superoxide (O_2^-) and other AOS such as H_2O_2 and OH^- took place. In addition to this oxidative burst, the cells showed increased lipid peroxidation after 2 hours, followed by increases in extracellular peroxidases and phenolic compounds. Cell suspension cultures of tomato genotype Cf5 treated with AF containing AVR5 elicitor showed a quick increase in extracellular production of AOS which was accompanied by increase in H^+-ATPase activity causing acidification of the extracellular medium (Vera-Estrella *et al.*, 1994a, b).

The observed acidification of the culture medium of suspension cells of tomato genotype Cf5 upon addition of the AVR5 elicitor, as found by Vera-Estrella *et al.* (1994a, b), contrasts with responses to non-specific elicitors (Vera-Estrella *et al.*, 1993; Felix *et al.*, 1993; Nürnberger *et al.*, 1994; Ebel and Cosio, 1994) which involved, among others, a rapid alkalization of the extracellular medium. Alkalization of the medium was also found in tobacco cell suspensions treated with elicitins from *Phytophthora* species, such as cryptogein (Ricci *et al.*, 1993). The acidification as opposed to alkalization of the medium of cell suspension cultures after treatment with race-specific or nonspecific elicitors, respectively, suggests that resistance gene-mediated defense responses induced by race-specific elicitors differ from non-specific elicitor-induced defense responses.

Research on various other host–pathogen interactions has provided substantial additional support for the involvement of AOS in the early events following recognition of the pathogen (Sutherland, 1991; Atkinson, 1993; Tzeng and DeVay, 1993; Lamb, 1994). The general picture that emerges is that the generated AOS might damage the invading pathogen directly and could initiate host cell wall lignification reactions, whereas intracellular AOS will cause the oxidation of lipids present in the membranes of the reacting host cells. This oxidative membrane damage might result in leakage of electrolytes from the host cells and could initiate the process of HR. Recent experiments have revealed that the activation of genes encoding PR proteins in tobacco can be the result of an increase in the endogenous H_2O_2 concentration, caused by inhibition of H_2O_2-catalases (Chen et al., 1993).

Injection of the purified race-specific elicitors AVR4 and AVR9 of C. fulvum into leaflets of Cf4 or Cf9 tomato genotypes induced expression of several genes, encoding PR proteins (Wubben et al., 1995). In Cf4 genotypes the AVR4 peptide induced mRNAs encoding acidic 1,3-β-glucanase and acidic chitinase within 4 h after injection, reaching a maximum at 8 h. In Cf9 genotypes the AVR9 peptide induced a similar transient expression of the acidic chitinase, whereas the expression of the acidic 1,3-β-glucanase was induced between 8 and 16 h, reaching a maximum at 24 h. In Cf0 genotypes the PR mRNAs were hardly induced. The induction pattern of basic chitinase and basic 1,3-β-glucanase was not as clear. The wound response caused by the injection procedure most probably released ethylene that induced transient expression of both basic chitinase and 1,3-β-glucanase at 1 h after injection. The differences in induction of basic PR-proteins between AVR4-treated Cf0 and Cf4 genotypes were not significant. In AVR9-treated Cf9 genotypes a significant increase in basic chitinase and 1,3-β-glucanase mRNA levels occurred between 8 and 16 h, reaching a peak at 24 h. In leaves of heterozygous Cf2 and Cf9 plants treated with AF containing AVR2 or AVR9 elicitor, respectively, Ashfield et al. (1994) found comparable results for induction of acidic and basic 1,3-β-glucanases. The mRNAs of basic 1,3-β-glucanase appeared at 6 h and of acidic 1,3-β-glucanase at 12 h, and reached maximum levels at 12 and 48 h, respectively.

B. RESPONSES INDUCED BY NIP1 AND PARASITICEIN

It is assumed that R. secalis secretes the necrosis-inducing proteins NIP1, NIP2 and NIP3 during colonization of susceptible plants, and that these NIPs kill the host tissue and stimulate the release of nutrients from the cells (Wevelsiep et al., 1993). Like fusicoccin (Marrè, 1979), the NIPs stimulate H^+-ATPase which might cause their toxic effects.

Besides the activation of cell necrosis in barley, which is a fairly late response and not cultivar-specific, NIP1 specifically induces the accumulation of the PR protein PRHv-1 in the barley cultivar Atlas 46 which contains the resistance gene *Rrs1*. The induction of PRHv-1 by the NIP1 protein or an avirulent race occurs early in Atlas 46, whereas induction of necrosis does not occur. Apparently growth of an avirulent race in cultivar Atlas 46 is inhibited before induction of necrosis can occur.

Most defense responses studied with elicitins have been carried out with cryptogein (β-elicitin) and parasiticein (α-elicitin). Cryptogein induces more distal necrosis than parasiticein (Nespoulos *et al.*, 1992; Kamoun *et al.*, 1993). Both cyptogein and parasiticein induce the CHS8 promoter of bean measured by GUS activity in transgenic tobacco transformed with the *CHS8: GUS* onstruct. Cryptogein induces alkalization of the medium of tobacco suspension cells within 10 min, followed by an oxidative burst within 30 min and subsequently by ethylene production and capsidiol synthesis (Ricci *et al.*, 1993).

IV.   RECEPTORS FOR RACE-SPECIFIC ELICITORS OF *C. FULVUM*

In order to investigate whether plasma membrane-bound receptors are involved in the process of elicitor perception, plasma membranes were isolated from a suspension culture of Cf5 cells and incubated those with a crude AF preparation containing the AVR5 elicitor were incubated (Vera-Estrella *et al.*, 1994a, b). Immediately upon exposure of the membranes to the elicitor preparation, ATPase activity increased 4-fold whereas other AF preparations lacking AVR5 had no effect on the ATPase activity. The increase in ATPase activity was due to a stimulation of the proton pump (H^+-ATPase), since incubation of Cf5 cells with the AVR5 elicitor preparation induced an immediate acidification of the culture medium. Activation of the H^+-ATPase was caused by dephosphorylation of the enzyme as shown by radiolabeling experiments and studies with inhibitors of protein phosphatases. Inhibition of protein kinase activity did not interfere with the elicitor-induced increase in H^+-ATPase activity, indicating that protein kinases, if involved, are further down the transduction pathway. The activation of the H^+-ATPase might acidify the cell wall, decrease its rigidity, and change the distribution of different ions across the plasma membrane. These processes might, among other effects, stimulate the production of callose, which is deposited at penetration sites in incompatible tomato–*C. fulvum* interactions (Lazarovits and Higgins, 1975; De Wit, 1977). Upon addition of the AVR5 elicitor preparation to Cf5 tomato suspension cells, callose synthesis was observed immediately, indicating that this is a very early response after elicitor perception (Vera-Estrella *et al.*, 1994a, b).

Binding of the AVR5 elicitor presumably occurs to a receptor in the plasma membrane, which via a signal transduction cascade activates H^+-ATPase by dephosphorylation. Extensive research on signal transduction chains in mammals has resulted in the identification of a family of heterotrimeric guanine nucleotide (GDP or GTP)-binding regulatory proteins (G proteins) that are involved in transduction across the plasma membrane of stimuli generated by binding of a signal molecule to its corresponding receptor (Kaziro *et al.*, 1991; Hepler and Gilman, 1992). Incubation of Cf5 plasma membranes with the AVR5 elicitor preparation and guanidine nucleotide analogs that inactivate or activate the α-subunit of a G protein, indicated that *Cf5*-mediated signaling responses take place via G proteins (Vera-Estrella *et al.*, 1994a, b).

Results of Vera-Estrella *et al.* (1994a, b) indicate that specific receptors for the *C. fulvum*-encoded AVR5 elicitor are present in the plasma membrane of the host cells. Studies by M. Kooman-Germann *et al.* (unpublished) showed that labeling of the AVR9 elicitor with ^{125}I does not affect its specific necrosis-inducing activity on Cf9 genotypes. Presently, binding studies using ^{125}I-labeled AVR9 elicitor and plasma membrane fractions of different tomato cultivars are being carried out by De Wit's research group (M. Kooman-Gersmann *et al.*, unpublished). The establishment of conditions that allow specific binding of a race-specific elicitor to its corresponding receptor might enable isolation of the receptor by affinity chromatography. In this way sufficient receptor protein might be isolated for N-terminal and internal sequencing. Amino acid sequence information on the receptor proteins allows the design of degenerated oligonucleotide probes to screen a cDNA or genomic library of tomato and eventually cloning of the encoding gene.

Preliminary results obtained with binding experiments using ^{125}I-labeled AVR9 elicitor and plasma membrane fractions of Cf9 and Cf0 tomato genotypes revealed no significant differences in binding properties between membranes of both cultivars (M. Kooman-Gersmann *et al.*, unpublished). This finding raises the question of how the *Cf9* resistance gene and the AVR9 receptor relate to each other. The results suggest that binding of the elicitor to membranes occurs in both resistant and susceptible genotypes, but that only in the resistant genotypes is a signal transduction cascade activated leading to a resistance response.

Genes encoding plasma membrane-bound receptors of elicitors of *C. fulvum* might also be isolated by exploiting the possible homology between these receptors and receptors that have already been identified in mammalian and plant systems (Hahn, 1995). Conserved regions of amino acids, identified in certain classes of receptor proteins, can form the basis to design degenerated primers that are used in PCR reactions on genomic DNA of the various tomato genotypes. Generated fragments can be analyzed for the presence of sequences encoding parts of a putative plasma membrane-bound receptor, and can be used to screen a genomic or cDNA library of tomato.

V. FUNGAL RESISTANCE GENES IN PLANTS

Little is known about receptors in the host plant, discussed above, corresponding to race-specific elicitors encoded by avirulence genes. Are the receptors indeed the products of resistance genes, as is assumed in the models proposed for gene-for-gene relationships, or do resistance genes encode dominant mutant proteins that are involved in signal transduction pathways which, when activated by the corresponding ligands, result in a programmed cell death? In the latter case a resistance gene can encode any protein of such a signal transduction pathway between receptor and HR response. Recently one viral (Whitham *et al.*, 1994), two bacterial (Martin *et al.*, 1993; Bent *et al.*, 1994; Mindrinos *et al.*, 1994), and two fungal resistance genes (Jones *et al.*, 1994; Lawrence *et al.*, 1994) have been cloned.

Jones and co-workers (Jones *et al.*, 1994; Hammond-Kosack and Jones, 1994) have employed two strategies to clone fungal resistance genes. Their efforts were focused mainly on cloning *Cf* genes of tomato that confer resistance to particular races of *C. fulvum*. The *Cf2* gene has been isolated by map-based cloning and the *Cf9* gene by transposon tagging. As the *Cf9* resistance gene is the only fungal resistance gene cloned so far of which the complementary AVR9 race-specific elicitor is fully characterized, I will limit the discussion mainly to the structure and potential properties of this gene and compare it with the cloned viral, bacterial, and other fungal resistance genes. A summary of cloned resistance genes and their products is presented in Table II.

A. MAP-BASED CLONING OF RESISTANCE GENES

Map-based cloning of resistance genes is straightforward, reliable, and the best way to proceed when there is lack of knowledge about the products of the resistance genes and about their complementary avirulence genes (Ellis *et al.*, 1988). The latter is true for nearly all documented gene-for-gene models. The availability of cloned avirulence genes and their products has usually not been helpful in cloning the corresponding resistance genes in the plant. This is particularly true when the product of an avirulence gene does not induce a clear HR response in the resistant plant. Nearly all bacterial avirulence genes cloned so far do not encode an HR-inducing race-specific elicitor (Keen, 1992), whereas the two cloned avirulence genes of *C. fulvum* and the putative avirulence genes of *R. secalis* and *P. parasitica* do.

The map-based cloning strategy has been successfully developed for plants that have been genetically well characterized, such as tomato (*Lycopersicon esculentum*), flax (*Linum usitatissimum*), and *Arabidopsis thaliana*. The bacterial resistance genes *Pto* of tomato and *RPS2* of *A. thaliana* have been cloned by the map-based approach (Martin *et al.*, 1993; Bent *et al.*, 1994; Mindrinos *et al.*, 1994). Jones and co-workers have used map-based cloning

TABLE II

Homologies between the CF9 protein and proteins encoded by other plant resistance genes, and other eukaryotic genes

Protein	Domains present	Location	Ligand	Homology	Reference
Resistance gene products					
CF9	LRRs (28 repeats)[a]	TM?[b]	AVR9?	....	Jones et al. (1994)
N	LRRs (4 repeats) P-loop, Leucine zipper	CP[c]	TMV elicitor? Replicase? Coat protein	LRRs	Whitham et al. (1994)
RPS2	LRRs (14 repeats) P-loop, Leucine zipper	CP/TM?	Bacterial elicitor?	LRRs	Bent et al. (1994) Mindrinos et al. (1994)
L[6],	LRRs (2 repeats) P-loop	CP	Fungal elicitor?	LRRs	Lawrence et al. (1994)
PTO	Proteine Kinase	CP/PM[d]?	Bacterial elicitor?	no homology	Martin et al. (1993)
Other proteins					
PGIPs	LRRs (10 repeats)	EC[e]/PM?	Fungal polygalacturonase	LPRs + other[f]	Toubart et al. (1992)
RLK5	LRRs (21 repeats)	TM	Unknown	LRRs	Walker (1993)
TMK1	LRRs (13 repeats)	TM	Unknown	LRRs	Chang et al. (1992)
TMKL1	LRRs (7 repeats)	TM	Unknown	LPRs	Valon et al. (1993)
TOLL	LRRs (15 repeats)	TM	Activated spätzle protein	LRRs	Hashimoto et al. (1988)
GP 1bα	LRRs (7 repeats)	TM	Activated von Willebrand factor	LRRs	Titani et al. (1987)
PRF	LRRs (9 repeats)	CP/TM?	Bacterial elicitor?	LRRs	Salmeron et al. (1994a, b)

[a] LRRs, Leucine-rich repeats
[b] TM, Transmembrane
[c] CP, Cytoplasmic
[d] PM, Plasma membrane-bound
[e] EC, Extracellular
[f] Other, in addition to LRRs, domain B and C are homologous

to isolate the resistance gene *Cf2*. The *Cf*-resistance genes are clustered. *Cf4* and *Cf9* are closely linked and located on the short arm of chromosome 1 (Van der Beek *et al.*, 1992; Jones *et al.*, 1993) while *Cf2* and *Cf5* are allelic and located on chromosome 6 (Dickinson *et al.*, 1993; Jones *et al.*, 1993). For both clusters of genes a high resolution map of the surrounding regions has been generated by restriction fragment length polymorphism (RFLP), random amplified polymorphic DNA (RAPD) and amplified fragment length polymorphism (AFLP) markers in crosses of tomato plants segregating for *Cf* genes in the offspring (Van der Beek *et al.*, 1992; Dickinson *et al.*, 1993; Jones *et al.* 1993; Hammond-Kosack and Jones, 1995). The generation of numerous molecular markers in the vicinity of the various resistance genes, combined with phenotypic markers closely linked to the *Cf* genes, allowed rapid screening of recombinants and made genomic walking to the *Cf* gene clusters and subsequent cloning of individual *Cf* genes possible. Markers absolutely linked to resistance gene *Cf2* have been used to isolate cosmid clones from a library prepared from plants homozygous for *Cf2*. Positively hybridizing clones were used to transform a Cf0 tomato genotype (without *Cf* resistance genes). Some clones were able to generate transgenic tomato plants that had become resistant to races of *C. fulvum* carrying avirulence gene *Avr2*, indicating that these cosmid clones carry resistance gene *Cf2* (Hammond-Kosack and Jones, 1994; M. S. Dixon *et al.*, unpublished).

Many more genes will be cloned soon by this strategy. To date, none of the resistance genes corresponding to the isolated fungal avirulence genes discussed above have been cloned by the map-based approach.

B.   TRANSPOSON TAGGING OF RESISTANCE GENES

Transposon tagging involves the inactivation of a resistance gene by insertion of a transposable element (Ellis *et al.*, 1988). As most resistance genes are dominant, insertion would lead to a loss of function. It is, however, important to realize that this strategy works when only one resistance gene is to be inactivated by the transposon. Insertion occurs at such a low frequency that inactivation of two genes by transposon insertion in one individual is virtually impossible. As it has been found that transposons such as the maize transposable element *Ac* transpose to linked sites on the chromosome, it is useful to have a transposable element positioned closely linked to the resistance gene to be inactivated (Carroll *et al.*, 1995).

1.   Transposon tagging of C. fulvum, *resistance gene Cf9*
The maize transposable elements *Ac (Activator)* and *Ds (Dissociation)* are active in the heterologous host tomato and preferentially transpose to linked sites (Carrol *et al.*, 1995). The two elements were used to tag the resistance gene *Cf9* in tomato (Hammond-Kosack and Jones, 1995; Jones *et al.*, 1994).

If the non-autonomous *Ds* element is positioned at a site linked to the targeted gene, it remains silent until it is transactivated by a transposase in *trans* arrangement; this can be achieved by stabilized *Ac (sAc)* which provides transposase but is unable to transpose itself, as one of its border sequences is removed. For the tagging experiment Jones *et al.* (1994) used homozygous Cf9 plants carrying *Ds* linked to *Cf9* and *sAc*. The The *sAc* element is able to transactivate the *Ds* element, which is excised and integrated at nearby sites on the chromosome. As *Ds* preferentially transposes to linked sites, a targeted tagging of the *Cf9* gene should be initiated from a stock of plants that contains the *Ds* element closely linked to the resistance gene *Cf9*. As the transposition frequency is not high, a strong selection system for transposition of *Ds* into *Cf9* is very helpful. The *Avr9* gene of *C. fulvum* encoding the HR-inducing AVR9 peptide has been exploited in the screening for transposition events. The fact that expression of the *Avr9* gene is lethal in a Cf9 background provides a strong selection for mutants in the *Cf9* gene that have been inactivated by *Ds* insertion. Insertion of *Ds* into the *Cf9* gene can be identified by crossing the tagged plants with Cf0 plants homozygous for the *Avr9* gene, so that all the progeny would receive both *Cf9* and *Avr9* and should die. Seeds grown from those F1 plants die after the first leaves have been formed, unless the *Cf9* gene has been inactivated by *Ds* insertion. In the latter case the plants survive. The F1 with the tagged *Cf9* should normally be viable or might develop necrotic sectors due to revertants caused by re-excision of *Ds* in plants which still contain *sAc*. In this way it is possible to screen efficiently large numbers of F1 plants on *Cf9/Avr9*-mediated responses and to tag and eventually clone the *Cf9* resistance gene by plasmid rescue (Rommens *et al.*, 1992).

Resistance gene *Cf9* has recently been cloned by this strategy (Jones *et al.*, 1994). The *Cf9* resistance gene encodes a protein of 863 amino acids (Jones *et al.*, 1994). Seven domains could be designated. The N-terminal domain A of 23 amino acids is consistent with a signal peptide; domain B of 68 amino acids is cysteine-rich; domain C contains 28 imperfect repeats of a conserved 24 amino acid leucine-rich repeat (LRR); domain D contains 28 amino acids; domain E contains 18 amino acids and is very acidic; domain F contains 37 amino acids and is a presumed transmembrane domain; C-terminal domain G contains 21 amino acids, is very basic and concludes with the amino acids KKRY (K = lysine; R = arginine; Y = tyrosine). Twenty-two N-glycosylation sites are distributed between domains B, C and D. The overall structure of the CF9 protein is consistent with a glycoprotein of which the main part is extracellular with LRRs and only the C-terminal domain G is cytoplasmic. However, there is still the possibility that the protein is intracellular as the KKRY sequence could function as a signal for retrieval of membrane-bound proteins from the Golgi to the endoplasmic reticulum (Townsley and Pelham, 1994). The sequence information on resistance gene *Cf9* predicts that the *Cf9* gene is a member of a superfamily of LRR class of proteins. The LRR domain is present in many proteins, extracellular or cytoplasmic, and is thought to mediate specific protein–protein interactions (Table II; Suzuki *et al.*, 1990; Braun *et al.*, 1991; Kobe and Deisenhofer, 1993). It is presently not known which role(s) LRRs play in the CF9 protein. Are they involved in AVR9 binding or do they interact with other components of the signal transduction pathway?

2. Homology of the Cf9 gene with other members of the Cf-cluster

When the *Cf9* gene is used as a probe on a Southern blot of isogenic lines of tomato, including Cf9 and Cf0, there are at least 11 hybridizing bands in the Cf9 line and five in the Cf0 (Jones *et al.*, 1994). Three of the bands in the Cf9 line cosegregate with the *Cf9* gene, whereas the others were linked either distally or proximally. Therefore the *Cf9* gene seems to be a member of a clustered gene family. In this respect the *Cf9* gene resembles gene clusters reported for *Pto* (Martin *et al.*, 1993), L^6 (Lawrence *et al.*, 1994), and *Rp* (Pryor and Ellis, 1993). When the Cf9 probe was hybridized to RNA isolated from the Cf9 and the Cf0 line, single bands of about 3 kb were found in both lines indicating that Cf0 lines contain alleles that express mRNAs that are very homologous to Cf9 mRNAs. Therefore the molecular basis of the specificity remains unclear since both lines presumably produce very similar CF9-like proteins. This could be expected based on results obtained by M. Kooman-Gersmann *et al.* (unpublished), who found that the AVR9 elicitor molecule binds equally well to plasma membranes of Cf0 and Cf9 lines. As membranes of both Cf9 and Cf0 lines bind AVR9 peptide, there is the possibility that the CF9 and the CF0 proteins represent two-component membrane-bound proteins of which the LRRs of the extracellular domain bind AVR9 peptide and the cytoplasmic domain transduces a signal onto neighboring proteins and confers the specificity of the response. In this model the binding domain of the CF9 and CF0 protein would be similar and the transducing domain of the two proteins would be different. Another possibility is that the LRRs of the CF9 protein do not bind the AVR9 peptide but interact with other components of the signal transduction pathway. The temperature dependence of AVR9 binding to plasma membranes may indicate that the CF9 protein is a member of a complex of proteins involved in signal transduction (M. Kooman-Gersmann *et al.*, unpublished).

3. Homology between the CF9 protein and other proteins including proteins encoded by other resistance genes

The CF9 protein shows the highest homology to two other plant proteins containing LRRs, including the receptor-like protein kinases (RLPKs) from *Arabidopsis thaliana* (Walker, 1993) and the antifungal polygalacturonase-inhibiting proteins (PGIPs) from bean (Toubart *et al.*, 1992; De Lorenzo *et al.*, 1994), pear (Stotz *et al.*, 1993) and tomato (Stotz *et al.*, 1994). The homology is mainly in the LRRs, but the PGIPs also show much homology with domains B and D of the CF9 protein. The reported PGIPs are extracellular proteins, but may have membrane-bound homologs (G. De Lorenzo, personal communication).

Recently the *Melampsora lini* resistance gene L^6 from flax (Lawrence *et al.*, 1994), the *Pseudomonas syringae* resistance gene *RPS2* from *Arabidopsis thaliana* (Bent *et al.*, 1994; Mindrinos *et al.*, 1994) and the tobacco mosaic virus (TMV) resistance gene *N* of tobacco (Whitham *et al.*, 1994) have been cloned. All three resistance genes contain LRRs and ATP/GTP binding sites (P-loops). In addition, *N* and *RPS2* contain a leucine zipper supposed to be involved in protein–protein interactions. L^6 and *N* are cytoplasmic, whereas *RPS2* may be transmembrane or cytoplasmic. At first sight the *N*, L^6, and *RPS2* genes belong to a similar class of resistance genes, different from the *Cf9* resistance gene, which contains LRRs but lacks a P-loop and leucine zipper, and also different from the *Pseudomonas syringae* pv. *tomato* resistance gene *Pto* from tomato (Martin *et al.*, 1993), which is a cytoplasmic serine–threonine protein kinase with a putative membrane anchor. A summary of the homologies between the CF9 protein and products of other resistance genes cloned so far is presented in Table II.

A few plant genes which encode putatively cooperating proteins involved in ligand binding, which might also be the case for the *Cf9* resistance gene cluster, are schematically presented in Fig. 1. These include the *SLG/SRK* (*S*-Locus specific *G*lycoprotein/*S*-locus *R*eceptor *K*inase) gene pair involved in sporophytic incompatibility in *Brassicaceae* (Nasrallah and Nasrallah, 1993; Nasrallah *et al.*, 1994), the *Pto/Prf* cluster involved in *Pseudomonas syringae* pv. *tomato* resistance, and *Pseudomonas* resistance and *f*enthion sensitivity in tomato (Salmeron *et al.*, 1994a, b), and the *Cf9* cluster (Jones *et al.*, 1994). The general picture that emerges from the work on cloned resistance genes shows that they have much more in common than the numerous bacterial and fungal avirulence genes cloned to date (Keen, 1992; this chapter). Presumably the resistance gene products provide sufficient variation to recognize a variety of plant pathogens. The present data suggest that there are common or at least similar signal transduction pathways in plants leading to resistance which can be turned on by different classes of ligands originating from diverse pathogens such as viruses, bacteria, and fungi. As the *Avr9/Cf9* interaction is the only avirulence gene–resistance gene combination of which both gene products are identified, it will soon become an ideal model for detailed studies on signal perception and signal transduction.

4.   *Is the CF9 protein involved in AVR9 elicitor perception?*

From the data available it is unclear how the CF9 protein, if involved at all, participates in perception of the AVR9 elicitor or in transduction of a signal leading to HR and resistance. This preliminary conclusion is based on results obtained with Southern and northern blots containing DNA and RNA from Cf9 and Cf0 plants, respectively, probed with cDNA of the *Cf9* gene (Jones *et al.*, 1994), and from results obtained with binding studies of labeled AVR9 to plasma membranes of Cf0 and Cf9 plants (M. Kooman-Gersmann *et al.*,

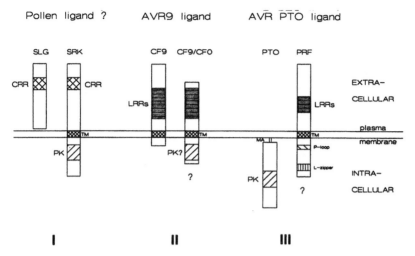

Fig. 1. Three pairs of putative bifunctional receptors for ligands derived from pollen (ligand unknown, I), *Cladosporium fulvum* (AVR9, II), and *Pseudomonas syringae* pv. *tomato* (AVRPTO, III). All pairs of bifunctional receptor genes presented are allelic or very closely linked. SLG, S-locus specific glycoprotein; SRK, S-locus receptor kinase; CRR, cysteine-rich region; LRRs, leucine-rich repeats; CF9, proteins encoded by resistence gene *Cf9*; CF9/0, proteins encoded by *Cf9/Cf0* alleles lacking functional *Cf9* gene; PRF, *Pseudomonas* resistance and *f*enthion sensitivity; TM, transmembrane; MA, membrane anchor; PK, protein kinase domain; P-loop: ATP/GTP binding site; L-zipper: leucine zipper; ?, hypothetical.

unpublished). This raises the question as to what are the other genes. Are they non-functional copies of the *Cf9* gene generated by unequal cross-over during meiosis, or do they represent homologous gene(s) encoding the functional receptor(s) for the AVR9 elicitor which after AVR9 perception interact with the CF9 protein? This could explain why there are functional receptors present in plasma membranes of both cultivars which bind the AVR9 peptide equally well. As the signal sequence KKRY present in the CF9 protein (Jones *et al.*, 1994), could function as a signal for retrieval of membrane-bound proteins from the Golgi to the endoplasmic reticulum (Townsley and Pelham, 1994), it is still possible that the CF9 protein is cytoplasmic. In this case the CF9 protein is not involved in AVR9 perception since binding studies support that the AVR9 receptor is plasma membrane-bound (M. Kooman-Gersmann *et al.*, unpublished). This would favour the possibility that only proteins encoded by Cf9/Cf0 alleles are involved in AVR9 perception. Another possibility is that PGIP-like proteins are involved in perception of the AVR9 elicitor. However, PGIP does not map at the *Cf9* locus (D. A. Jones, unpublished), and is in this respect different from the gene pairs such as *SLG/SRK*, *Pto/Prf* and

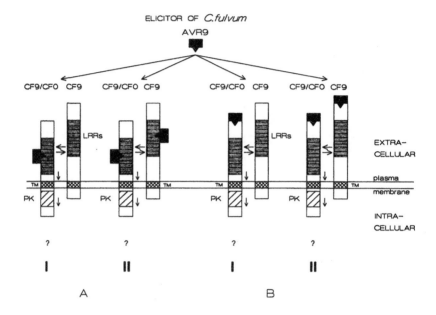

Fig. 2.   Possible models in which the AVR9 elicitor of *C. fulvum* interacts with putative receptors in Cf9 and Cf0 genotypes. The models are based on the following facts: (a) the amino acid sequence of the CF9 protein is known; most probably the LRRs of the CF9 protein are extracellular as is proposed in models I and II; (b) plasma membranes of tomato genotypes Cf9 and Cf0 bind AVR9 elicitor equally well. In (A) LRRs are involved in AVR9 perception; in (B) LRRs do not interact with the AVR9 elicitor but rather modulate the CF9 protein after AVR9 binding. For abbreviations, see Fig. 1. (A) I: AVR9 peptide binds to LRRs of proteins (transmembrane) encoded by *Cf9/Cf0* alleles, followed by interaction with the CF9 protein (transmembrane) and signaling. II: AVR9 peptide binds to LRRs of proteins (transmembrane) encoded by both *Cf9/Cf0* alleles and the CF9 protein (transmembrane), followed by modulating of the two proteins and signaling. (B) Similar to (A) except that domains of the proteins other than LRRs are assumed to be involved in perception of the AVR9 elicitor. After perception modulation and further signaling through other membrane-bound and cytoplasmic proteins follows.

Cf9/Cf0, discussed before. Due to the strong homology between the PGIP protein and the CF9 protein, involvement of PGIP in AVR9 binding or modulating proteins involved in AVR9 binding should not be ruled out. The different scenarios for AVR9 perception and signal transduction are schematically presented in Fig. 2.

VI. CONCLUSIONS AND FUTURE PROSPECTS FOR EXPLOITATION OF AVIRULENCE GENES AND RESISTANCE GENES

The interaction between tomato and *C. fulvum* has proven to be an excellent model system to dissect a host–pathogen interaction that complies with the gene-for-gene model at the molecular level. Cloned avirulence and resistance genes will provide valuable tools to study the molecular genetic mechanisms determining race-specific resistance. Once sufficient amounts of the avirulence gene products, the race-specific elicitors, are available, the biochemical and cellular mechanisms in plants and plant cells eventually leading to resistance can be studied in detail. Leaves as well as suspension cells can be treated with different concentrations of these elicitors to follow early events in signal transduction pathways.

Detailed information on the mode of action of specific elicitors might be obtained by structure–function analysis of the elicitor proteins that specifically interact with receptors, an approach currently followed by R. Vogelsang *et al.* (unpublished). By PCR-mediated site-directed mutagenesis, specific mutations will be introduced in the ORFs of the *Avr4* and *Avr9* genes and the resulting modified elicitors will be tested for their ability to induce the HR in tomato genotypes Cf4 and Cf9, respectively. ^{1}H NMR and crystallographic studies on AVR-proteins will provide further information on their molecular structure. Detailed information on the mechanism that underlies the specific interaction between elicitors of *C. fulvum* and host-encoded receptors might form the basis for elucidation of the signal perception and transduction pathways in plant-pathogen interactions.

Probes for avirulence genes will facilitate studies on the distribution of these genes in populations of the same and related fungal species. These studies could clarify the molecular basis of variation and adaptation to host plants. From the avirulence genes cloned so far we know that fungal pathogens have adapted three strategies to avoid recognition by their hosts. These include deletions, point mutations, and insertions in their avirulence genes. Understanding the mechanisms employed by fungal plant pathogens to avoid recognition by the host, could be instrumental in developing new strategies for crop protection and breeding for disease resistance.

Combining an avirulence gene and the complementary resistance gene in host plants under control of fungus-inducible promoters of either gene could possibly give new leads for molecular resistance breeding (two-component sensor system; De Wit, 1992). Activation of the avirulence gene by pathogens, which should occur both quickly and locally, could result in production of the race-specific elicitor that interacts with the resistance gene product. As a result of this interaction, a localized HR will be induced which will prevent further spread of any invading pathogen that can be inhibited by the HR, followed by a general defense response. With this system it should be possible to use a highly specific resistance gene/avirulence gene combination (for example *Cf9/Avr9*) in combination with a general pathogen-inducible promoter to obtain plants resistant against diverse pathogens. Preliminary results with tomato plants containing the two-component system look promising (G. Honée *et al.*, unpublished results). For applications in plants other than tomato, the complementary resistance gene *Cf9* is needed. As this has been cloned recently by Jones *et al.* (1994), in the near future the feasibility of this approach can be tested in crop plants in which the *Cf9* gene can be introduced by transformation.

Detailed information on the molecular structure of disease resistance genes has become available. Some encode receptor-like proteins with LRRs, P-loops and leucine zippers, some encode receptor-like proteins with LRRs and P-loops, some with LRRs only, and one encodes a serine/threonine protein kinase. These general structures suggest that they all function in signal transduction cascades where protein–protein interactions and (de)phosphorylation events take place. Some proteins contain membrane-spanning and extracellular domains, enabling possible interactions with ligands such as race specific-elicitors. Positive or negative proofs for direct interactions between race-specific elicitors and resistance genes products are still missing, but will become available in the near future.

The resistance genes cloned so far share many features. They could form the basis for the molecular design of synthetic resistance genes able to recognize or to respond to not just one isolate of a pathogen but to various pathogens. These synthetic genes should be transferred to different economically important crop species in order to protect them against these pathogens.

ACKNOWLEDGEMENTS

I thank Dr Jonathan D. G. Jones, Dr Barbara Valent, Dr Wolfgang Knogge and Dr Sophien Kamoun for sending manuscripts prior to publication and my colleagues Guy Honée, Matthieu H. A. J. Joosten, Miriam Kooman-Gersmann and Ralph Vogelsang for critically reading the manuscript. I thank Paul Vossen for making the Figures.

REFERENCES

Ashfield, T., Hammond-Kosack, K. E., Harrison, K. and Jones, J. D. G. (1994). *Cf* gene dependent induction of 1,3-β-glucanase promoter in tomato plants infected with *Cladosporium fulvum*. *Molecular Plant–Microbe Interactions* **7**, 645–656.

Atkinson, M. M. (1993). Molecular mechanisms of pathogen recognition by plants. *Advances in Plant Pathology* **10**, 35–64.

Bent, A. F., Kunkel, B. N., Dahlbeck, D., Brown, K. L., Schmidt, R., Giraudat, J., Leung, J. and Staskawicz, B. J. (1994). *RPS2* of *Arabidopsis thaliana*: a leucine-rich repeat class of plant disease resistance genes. *Science* **265**, 1856–1860.

Biffen, R. H. (1905). Mendel's laws of inheritance and wheat breeding. *Journal of Agricultural Science* 1, 4–48.

Braun, T., Schofield, P. R. and Sprengel, R. (1991) Amino-terminal leucine-rich repeats in gonadotropin receptors determine hormone selectivity. *EMBO Journal* 10, 1886–1890.

Caddick, M. X. (1992). Characterization of a major *Aspergillus* regulatory gene *are*A. *In* "Molecular Biology of Filamentous Fungi" (U. Stahl and P. Tudzynski, eds), pp. 141–152. VCH, Weinheim.

Carroll, B. B., Klimyuk, V. I., Thomas, C. M., Bishop, G. J. and Jones J. D. G. (1995). Germinal transposition of the maize element *Dissociation* from T-DNA loci in tomato. *Genetics* 139, 407–420.

Chang, C., Schaller, G. E., Patterson, S. E., Kwok, S. F., Meyerowitz, E. M. and Bleecker, A. B. (1992). The *TMK1* gene from *Arabidopsis* codes for a protein with structural and biochemical characteristics of a receptor protein kinase. *The Plant Cell* 5, 371–378.

Chen, Z., Silva, H. and Klessig, D. F. (1993). Active oxygen species in the induction of plant systemic acquired resistance by salicylic acid. *Science* 262, 1883–1886.

Crute, I. R. (1985). The genetic bases of relationships between microbial parasites and their hosts. *In* "Mechanisms of Resistance to Plant Diseases" (R. S. S. Fraser, ed.), pp. 80–42. Nijhoff/Junk, Dordrecht.

Daboussi, M. J., Langin, T., and Brigoo, Y. (1992). Fot1, a new class of fungal transposable elements. *Molecular and General Genetic* 232, 12–16.

Danhash, N., Wagemakers, C. A. M., Van Kan, J. A. L. and De Wit, P. J. G. M. (1993). Molecular characterization of four chitinase cDNAs obtained from *Cladosporium fulvum*-infected tomato. *Plant Molecular Biology* 22, 1017–1029.

De Lorenzo, G., Cervone, F., Bellincampi, D., Caprari, C., Clark, A. J., Desiderio, A., Forrest, L., Leckie, F., Nuss, L. and Salvi G. (1994). Polygalacturonase, PGIP and oligogalacturonides in cell–cell communication. *Biochemical Society Transactions* 22, 394–397.

De Wit, P. J. G. M. (1977). A light and scanning electron-microscopic study of infection of tomato plants by virulent and avirulent races of *Cladosporium fulvum*. *Netherlands Journal of Plant Pathology* 83, 109–122.

De Wit, P. J. G. M. (1986). Elicitation of active resistance mechanisms. *In* "Biology and Molecular Biology of Plant–Pathogen interactions", NATO ASI series: Volume H1 (J. Bailey, ed.), pp 149–169. Springer-Verlag, Berlin.

De Wit, P. J. G. M. (1992). Molecular characterization of gene-for-gene systems in plant–fungus interactions and the application of avirulence genes in control of plant pathogens. *Annual Review of Phytopathology* 30, 391–418.

De Wit, P. J. G. M. and Kodde, E. M. (1981). Induction of polyacetylenic phytoalexins in *Lycopersicon esculentum* after inoculation with *Cladosporium fulvum* (syn. *Fulvia fulva*). *Physiological Plant Pathology* 18, 297–314.

De Wit, P. J. G. M. and Spikman, G. (1982). Evidence for the occurrence of race-and cultivar-specific elicitors of necrosis in intercellular fluids of compatible interactions between *Cladosporium fulvum* and tomato. *Physiological Plant Pathology* 21, 1–11.

De Wit, P. J. C. M and Van der Meer, F. E. (1986). Accumulation of the pathogenesis-related tomato leaf protein P14 as an early indicator of incompatibility in the interaction between *Cladosporium fulvum* (syn. *Fulvia fulva*) and tomato. *Physiological and Molecular Plant Pathology* 28, 203–214.

De Wit, P. J. G. M., Hofman, J. E. and Aarts, J. M. M. J. G. (1984). Origin of specific elicitors of chlorosis and necrosis occurring in intercellular fluids of compatible interactions of *Cladosporium fulvum* (syn. *Fulvia fulva*) and tomato. *Physiological Plant Pathology* 24, 17–23.

De Wit, P. J. G. M., Buurlage, M. B. and Hammond, K. E. (1986). The occurrence of host, pathogen and interaction-specific proteins in the apoplast of *Clado-sporium fulvum* (syn. *Fulvia fulva*) infected tomato leaves. *Physiological and Molecular Plant Pathology* **29**, 159-172.

De Wit, P. J. G. M., Toma, I. M. J. and Joosten, M. H. A. J. (1988). Race-specific elicitors and pathogenicity factors in the *Cladosporium fulvum*-tomato interaction. *In* "Physiology and Biochemistry of Plant–Microbial Interactions" (N. T. Keen, T. Kosugesand L. L. Walling, eds), pp. 111-119. The American Society of Plant Physiologists, Rockville MD, USA.

De Wit, P. J. G. M., Hofman, J. E., Velthuis, G. C. M. and Kuc, J. A. (1985). Isolation and characterization of an elicitor of necrosis isolated from intercellular fluids of compatible interactions of *Cladosporium fulvum* (syn. *Fulvia fulva*) and tomato. *Plant Physiology* **77**, 642-647.

Dickinson, M. J., Jones, D. A. and Jones, J. D. G. (1993). Close linkage between the *Cf-2/Cf-5* and *Mi* resistance loci in tomato. *Molecular Plant–Microbe Interactions* **6**, 341-347.

Dietrich R. A., Delaney, T. P., Uknes, S. J., Ward, E. R., Ryals, J. A. and Dangl, J. L. (1994). *Arabidopsis* mutants simulating disease resistance. *Cell* **77**, 565-577.

Dixon, R. A., Harrison, M. J. and Lamb, C. J. (1994). Early events in the activation of plant defense responses. *Annual Review of Phytopathology* **32**, 479-501.

Dyck, P. L. and Schaller, C. W. (1961). Inheritance of resistance in barley to several physiologic races of the scald fungus. *Canadian Journal of Genetic Cytology* **3**, 153-164.

Ebel, J. and Cosio E. (1994). Elicitors of plant defense responses. *International Review of Cytology* **148**, 1-36.

Ellis, J. G., Lawrence, G. J., Peacock, W. J. and Pryor A. I. (1988). Approaches to cloning plant resistance genes conferring resistance to fungal pathogens. *Annual Review of Phytopathology* **26**, 245-263.

Farrer, W. (1898). The making and improvement of wheats for Australian conditions. *Agricultural Gaz NSW* **9**, 131-168.

Felix, G., Regenass, M. and Boller, B. (1993). Specific perception of subnanomolar concentrations of chitin fragments by tomato cells: induction of cellular alkalinization, changes in protein phosphorylation, and establishment of a refractory state. *The Plant Journal* **4**, 307-316.

Flor, H. H. (1942). Inheritance of pathogenicity in *Melampsora lini. Phytopathology* **32**, 653-669.

Fu, Y. H. and Marzluf, G. A. (1990). *Nit-2*, the major positive-acting nitrogen regulatory gene of *Neurospora crassa*, encodes a sequence-specific DNA-binding protein. *Proceedings National Academy of Sciences USA* **87**, 5331-5335.

Gabriel, D. W. and Rolfe, B. G. (1990). Working models of specific recognition in plant–microbe interactions. *Annual Review of Phytopathology*, **28**, 365-391.

Greenberg, J. T., Guo, A., Klessig, D. F. and Ausubel, F. M. (1994). Programmed cell death in plants: a pathogen-triggered response activated coordinately with multiple defense functions. *Cell*, **77**, 551-563.

Habgood, R. M. and Hayes, J. D. (1971). The inheritance of resistance to *Rhynchosporium secalis* in barley. *Heredity* **27**, 25-37.

Hahn, M. G. (1995). Signal perception at the plasma membrane: binding proteins and receptors. *In* "Membranes: Specialized Functions in Plants" (D. L. Smallwood, J. P. Knox and D. J. Bowles, eds), in press.

Hahn, M., Jüngling, S. and Knogge, W. (1993). Cultivar-specific elicitation of barley defense reactions by the phytotoxic peptide NIP1 from *Rhynchosporium secalis. Molecular Plant–Microbe Interactions* **6**, 745-754.

Hammond-Kosack, K. E. and Jones, J. D. G. (1995). Plant disease resistance genes, unravelling how they work. *Canadian Journal of Botany*, in press.

Hammond-Kosack, K. E., Harrison, K. and Jones, J. D. G. (1994). Developmentally regulated cell death on expression of the fungal avirulence gene *Avr9* in tomato seedlings carrying the disease-resistance gene *Cf9*. *Proceedings of National Academy of Sciences USA* **91**, 10445–10449

Hammond-Kosack, K. E., Staskawicz, B. J., Jones, J. D. G. and Baulcombe, D. C. (1995) Functional expression of a fungal avirulence gene from a "modified potato virus X genome. *Molecular Plant — Microbe Interactions* **8**, 181–185

Hashimoto, C., Hudson, K. and Anderson, K. V. (1988). The *Toll* gene of *Drosophila*, required for dorsal-ventrat embryonic polarity appears to encode a transmembrane protein. *Cell* **52**, 269–279.

Hepler, J. R. and Gilman, A. G. (1992). G proteins. *Trends In Biological Science* **17**, 383–387.

Huet, I.-C., Bouaziz, S., Nespoulos, C., Van Heijenoort, C., Guittet, E.. and Pernollet, J.-C. (1994). The phospholipase A_2 nature of elicitins shown by the 3D-structure and the catalytic properties of capsicein. *In* "Abstracts 4th International Congress of Plant Molecular Biology" Amsterdam June 19–24, 1994.

Jones, D. A., Dickinson, M. J., Balint-Kurti, P. J., Dixon, M. S. and Jones, J. D. G. (1993). Two complex resistance loci revealed in tomato by classical and RFLP mapping of the *Cf-2, Cf-4, Cf-5*, and *Cf-9* for resistance to *Cladosporium fulvum*. *Molecular Plant–Microbe Interactions* **6**, 348–357.

Jones, D. A., Thomas, C. M., Hammond-Kosack, K. E., Balint-Kurti, P.J. and Jones, J. D. G. (1994). Isolation of the tomato *CF-9* gene for resistance to *Cladosporium fulvum* by transposon tagging, *Science*, **266**, 789–793.

Joosten, M. H. A. J. and De Wit, P. J. G. M. (1989). Identification of several pathogenesis-related proteins in tomato leaves inoculated with *Cladosporium fulvum* (syn. *Fulvia fulva*) as 1,3-ß-glucanases and chitinases. *Plant Physiology* **89**, 945–951.

Joosten, M. H. A. J., Cozijnsen, A. J. and De Wit, P. J. G. M. (1994). Host resistance to a fungal tomato pathogen lost by a single base-pair change in an avirulence gene. *Nature* **367**, 384–387.

Joosten, M. H. A. J., Verbakel, M., Nettekoven, M. E., van Leeuwen, J., van der Vossen, R. T. M. and De Wit, P. J. G. M. (1995). The phytopathogenic fungus is not sensitive to the chitinase and 1,3-β-glucanase defence proteins of its host tomato. *Physiological and Molecular Plant Pathology* **46**, 45–59.

Kamoun, S., Young, M., Glascock, C. and Tyler, B. M. (1993a). Extracellular protein elicitors from *Phytophthora*: host specificity and induction of resistance to fungal and bacterial phytopathogens. *Molecular Plant–Microbe Interactions* **6**, 15–25.

Kamoun, S., Klucher, K. M., Coffey, M. D. and Tyler, B. M. (1993b). A gene encoding a host-specific elicitor protein of *Phytophthora parasitica*. *Molecular Plant–Microbe Interactions* **6**, 573–581.

Kamoun, S., Young, M. Foster, H., Coffey, M. D. and Tyler B. (1994) Potential role of elicitins in the interaction between *Phytophthora species* and tobacco. *Applied and Environmental Microbiology* **60**, 1593–1598.

Kaziro, Y., Itoh, H., Kozasa, T., Nakafuku, M. and Satoh, T. (1991). Structure and function of signal-transducing GTP-binding proteins. *Annual Review of Biochemistry* **60**, 349–400.

Keen, N. T. (1975) Specific elicitors of plant phytoalexin production: determinants of race-specificity in pathogens. *Science* **187**, 74–75.

Keen, N. T. (1982). Specific recognition in gene-for-gene host–parasite systems. *Advances in Plant Pathology* **1**, 35–82.

Keen, N. T. (1990). Gene-for-gene complementarity in plant–pathogen interactions. *Annual Review of Genetics* **24**, 447–463.

Keen, N. T. (1992). The molecular biology of disease resistance genes. *Plant Molecular Biology* **19**, 109–122.

Kiraly, Z. (1980). Defenses triggered by the invader: Hypersensitivity. *In* "Plant Disease", Volume 4 (J. G. Horsfall and E. B. Cowling eds), pp. 201–224. Academic Press, New York.

Knogge, W., Giertich, A., Hermann, H., Wernert, P. and Rohe M. (1995). Molecular identification of the *nip1* gene, an avirulence gene from the barley pathogen *Rhynchosporium secalis*. *In* "Advances in Molecular Genetics of Plant–Microbe Interactions, Current Plant Science and Biotechnology in Agriculture", Volume 3, (M. J. Daniels, ed.), 207–214. Kluwer Academic Publishers, Dordrecht.

Kobe, B. and Deisenhofer, J. (1993). Crystal structure of porcine ribonuclease inhibitor, a protein with leucine-rich repeats. *Nature* **366**, 751–756.

Lamb, C. J. (1994). Plant disease resistance genes in signal perception and transduction. *Cell* **76**, 419–422.

Lawrence, G. J., Ellis, J. G. and Finnegan, E. J. (1994). Cloning a rust-resistance gene in flax. *In* "Advances in Molecular Genetics of Plant–Microbe Interactions, Current Plant Science and Biotechnology in Agriculture", Volume 3, (M. J. Daniels, ed.), pp. 303–306. Kluwer Academic Publishers, Dordrecht.

Lazarovits, G. and Higgins, V. J. (1976). Ultrastructure of susceptible, resistant, and immune reactions of tomato to races of *Cladosporium fulvum*. *Canadian Journal of Botany* **54**, 235–247.

Legendre, L., Heinstein, P. F. and Low P. S. (1992). Evidence for participation of GTP-binding proteins in elicitation of the rapid oxidative burst in cultured soybean cells. *Journal of Biological Chemistry* **267**, 20140–20147.

Lehnackers, H. and Knogge, W. (1990). Cytological studies on the infection of barley cultivars with known resistance genotypes by *Rhynchosporium secalis*. *Canadian Journal of Botany* **68**, 1953–1963.

Long, S. R. and Staskawicz, B. J. (1993). Prokaryotic plant parasites. *Cell* **73**, 921–935.

Marmeisse R., Van den Ackerveken, G. F. J. M., Goosen, T., De Wit, P. J. G. M. and Van den Broek, H. W. J. (1993). Disruption of the avirulence gene *avr9* in two races of the tomato pathogen *Cladosporium fulvum* causes virulence on tomato genotypes with the complementary resistance gene *Cf9*. *Molecular Plant–Microbe Interactions* **6**, 412–417.

Marrè, E. (1979). Fusicoccin: a tool in plant physiology. *Annual Review of Plant Physiology* **30**, 273–288.

Martin, G. B., Brommonschenkel, S. H. Chunwongse, J., Frary, A., Ganal, M. W., Spivey, R., Wu, T., Earle, E. D. and Tanksley, S. D. (1993). Map-based cloning of a protein kinase gene conferring disease resistance in tomato. *Science* **262**, 1432–1436.

Mills, D. (1985). Transposon mutagenesis and its potential for studying virulence genes in plant pathogens. *Annual Review of Phytopathology* **23**, 297–320.

Mindrinos, M., Katagiri, F., Yu, G. L. and Ausubel F. M. (1994). The *A. thaliana* disease resistance gene *RPS2* encodes a protein containing a nucleotide-binding site and leucine-rich repeats. *Cell* **78**, 1089–1099.

Nasrallah, J. and Nasrallah, M. E. (1993). Pollen-stigma signaling in the sporophytic self-incompatibility response. *The Plant Cell* **5**, 1325–1335.

Nasrallah, J., Rundle S. J. and Nasrallah, M. E. (1994). Genetic evidence for requirement of the *Brassica S-* locus receptor kinase gene in the self-incompatibility response. *The Plant Journal* **5**, 373–384.

Nespoulos, C., Huet, J. C. and Pernollet, J. C. (1992). Structure–function relationships of α and β elicitins signal proteins involved in the plant–*Phytophthora* interaction. *Planta* **186**, 551–557.

Nürnberger, T., Nennstiel, D., Jabs, T., Sacks, W. R., Hahlbrock, K. and Scheel D. (1994). High affinity binding of a fungal oligopeptide elicitor to parsley plasma membranes triggers multiple defense responses. *Cell* 378, 449–460.

Oort, A. J. P. (1944). Onderzoekingen over stuifbrand. II. Overgevoeligheid voor stuifbrand (*Ustilago tritici*) with a summary: hypersensitiveness of wheat to loose smut. *Tijdschrift over Planteziekten* **50**, 73–106.

Peever, T. L. and Higgins, V. J. (1989). Electrolyte leakage, lipoxygenase, and lipid peroxidation induced in tomato leaf tissue by specific and nonspecific elicitors from *Cladosporium fulvum*. *Plant Physiology* **90**, 867–875.

Pernollet, J. C., Sailantin, M., Sallé-Tourne, M. and Huet, J. C. (1993). Elicitin isoforms from seven *Phytophthora* species: comparison of their physicochemical and topic properties to tobacco and other plant species *Physiological and Molecular Plant Pathology* **41**, 427–435.

Pryor, T. and Ellis, J. (1993). The genetic complexity of fungal resistance genes in plants. *Advances in Plant Pathology* **10**, 281–305.

Ricci, P., Bonnet, P., Huet, J. C., Sallantin, M., Beauvais-Cante F., Bruneteau, M., Billard, V., Michel, G. and Pernollet, J. C. (1989). Structure and activity of proteins from pathogenic fungi *Phytophthora* eliciting necrosis and acquired resistance in tobacco. *European Journal of Biochemistry* **183**, 555–563.

Ricci, P., Trentin, F., Bonnet, P., Venard, P., Mouton-Perronnet F. and Bruneteau, M. (1992). Differential production of parasiticein an elicitor of necrosis and resistance in tobacco, by isolates of *Phytophthora parasitica*. *Plant Pathology* **41**, 298–307.

Ricci, P., Panabieres, F., Bonnet, P., Maia, N., Ponchet, M., Devergne, J.-C., Marais, A., Cardin, L., Milat, M. L. and Blein, J. B. (1993). Proteinaceous elicitors of plant defense responses. *In* "Mechanisms of Plant Defense Responses", Volume 2 (B. Fritig and M. Legrand, eds), pp. 121–135. Kluwer Academic Publishers, Dordrecht, The Netherlands.

Rommens, C. M. T., Rudenko, G. N., Djikwel, P. P., van Haaren, M. J. J., Ouwerkerk, P. B. F., Blok, K. M., Nijkamp, H. J. J. and Hille J. (1992). Characterization of the Ac/Ds behaviour in transgenic tomato plants, using plasmid rescue. *Plant Molecular Biology* **20**, 61–70.

Salmeron, J. M., Barker, S. J., Carland, F. M., Metha, A .Y. and Staskawicz, B. J. (1994a). Tomato mutants altered in bacterial disease resistance provide evidence for a new locus controlling pathogen recognition. *The Plant Cell* **6**, 511–520.

Salmeron, J. M., Rommens, C., Barker, S. J., Carland, F. M. and Staskawicz, B. J. (1994b). Isolation of mutations in the *Pto* resistance locus of tomato and identification of a new locus, *Prf*, controlling pathogen recognition. Seventh International Symposium on Molecular Plant–Microbe Interactions, June 26th–July 1st, 1994. Edinburgh. Book of abstracts p. 86.

Scheel, D. (1990). Elicitor recognition and signal transduction in plant defense gene activation. *Zeitschrift fur Naturforschung* 45c, 569–575.

Scholtens-Toma, I. M. J. and De Wit, P. J. G. M. (1988). Purification and primary structure of a necrosis-inducing peptide from the apoplastic fluids of tomato infected with *Cladosporium fulvum* (syn. *Fulvia fulva*). *Physiological and Molecular Plant Pathology* **33**, 59–67.

Silué, D., Notteghem, J. L. and Tharreau, D. (1992a). Evidence for a gene-for-gene relationship in the *Oryza sativa–Magnaporthe grisea* pathosystem *Phytopathology* **82**, 577–580.

Silué, D., Tharreau, D. and Notteghem, J. L. (1992b). Identification of *Magnaporthe grisea* avirulence genes to seven rice cultivars. *Phytopathology* **82**, 1462–1467.

Stakman, E. C. (1917). Biologic forms of *Puccinia graminis* on cereals and grasses. *Journal of Agricultural Research* **10**, 429–495.

Stotz, U. H., Powell, A. L. T., Damon, S. E., Greve, L. C., Bennett, A. B. and Labavitch, J. M. (1993). Molecular characterization of a polygalacturonase inhibitor from *Pyrus communis* L. cv. Bartlett. *Plant Physiology* **102**, 113–138.

Stotz, U. H., Contos, J. J. A., Powell, A. L. T., Bennett, A. B. and Labavitch, J. M. (1994). Structure and expression of an inhibitor of fungal polygalacturonases from tomato. *Plant Molecular Biology* **25**, 607–617.

Sutherland, M. W. (1991). The generation of oxygen radicals during host plant responses to infection. *Physiological and Molecular Plant Pathology* **39**, 79–93.

Suzuki, N., Choe, H. R., Nishida, Y., Yamawaki-Kataoka, Y., Ohnishi, S., Tamaoki, T. and Kataoka, T. (1990). Leucine-rich repeats and carboxy terminus are required for interaction of yeast adenylate cyclase with RAS proteins. *Proceedings of the National of Sciences USA* **87**, 8711–8715.

Titani, K., Takio, K., Hada, M. Ruggeri, Z. M. (1987). Amino acid sequence of the von Willebrand factor-binding domain of a platelet membrane glycoprotein lb. *Proceedings of the National Academy of Sciences* **84**, 5610–5614.

Toubart, P., Desiderio, A., Salvi, G., Cervone, F., Daroda L., De Lorenzo, G., Bergmann, C., Darvill, A. G. and Albersheim, P. (1992). Cloning and characterization of the gene encoding the endopolygalacturonase-inhibiting protein (PGIP) of *Phaseolus vulgaris*. *The Plant Journal*. **2**, 367–373.

Townsley, F. M. and Pelham H. R. B. (1994). The KKXX signal mediates the retrieval of the membrane proteins from the Golgi to the ER in yeast. *European Journal of Cell Biology* **64**, 221–226.

Tzeng, D. D. and DeVay, J. E. (1993). Role of oxygen radicals in plant disease development. *Advances in Plant Pathology* **10**, 1–33.

Valent, B. and Chumley, F. G. (1991). Molecular genetic analysis of the rice blast fungus, *Magnaporthe grisea*. *Annual Review of Phytopathology* **29**, 443–467.

Valent, B. and Chumley, F. G. (1994). Avirulence genes and mechanisms of genetic instability in the rice blast fungus. *In* "Rice Blast Diseases" (R. S. Ziegler, S. A. Leong and P. S. Teng, eds). CAB International, Wallingford. pp. 111–134.

Valon, C., Smalle, J., Goodman, H. M. and Giraudat, J. (1993). Characterization of an *Arabidopsis thaliana* gene (*TMLK1*) encoding a putative transmembrane protein with an unusual kinase-like domain. *Plant Molecular Biology* **23**, 415–421.

Van den Ackerveken, G. F. J. M., Van Kan, J. A. L. and De Wit, P. J. G. M. (1992). Molecular analysis of the avirulence gene *avr9* of the fungal tomato pathogen *Cladosporium fulvum* fully supports the gene-for-gene hypothesis. *The Plant Journal* **2**, 359–366.

Van den Ackerveken, G. F. J. M., Vossen, J. P. M. J. and De Wit, P. J. G. M. (1993). The AVR9 race-specific elicitor of *Cladosporium fulvum* is processed by endogenous and plant proteases. *Plant Physiology* **103**, 91–96.

Van den Ackerveken, G. F. J. M., Dunn, R. M., Cozijnsen, A. J., Vossen, J. P. M. J., Van den Broek, H. W. J. and De Wit, P. J. G. M. (1994). Nitrogen limitation induces expression of the avirulence gene *avr9* in the tomato pathogen *Cladosporium fulvum*. *Molecular and General Genetics* **243**, 277–285.

Van der Beek, J. G., Verkerk, R., Zabel, P. and Lindhout, P. (1992). Mapping strategy for resistance genes in tomato based on RFLPs between cultivars: (resistance to *Cladosporium fulvum*) on chromosome 1. *Theoretical and Applied Genetics* **84**, 106–112.

Van Kan, J. A. L, Van den Ackerveken, G. F. J. M. and De Wit, P. J. G. M. (1991). Cloning and characterization of cDNA of avirulence gene *avr9* of the fungal tomato pathogen *Cladosporium fulvum*, causal agent of tomato leaf mold. *Molecular Plant–Microbe Interactions* **4**, 52–59.

Van Kan, J. A. L., Joosten, M. H. A. J., Wagemakers, C. A. M., Van den Berg-Velthuis, G. C. M. and De Wit, P. J. G. M. (1992). Differential accumulation of mRNAs encoding extracellular and intracellular PR proteins in tomato induced by virulent and avirulent races of *Cladosporium fulvum*. *Plant Molecular Biolology* **20**, 513–527.

Vera-Estrella, R., Blumwald, E. and Higgins, V. J. (1992). Effect of specific elicitors of *Cladosporium fulvum* on tomato suspension cells. *Plant Physiology* **99**, 1208–1215.

Vera-Estrella, R., Blumwald, E. and Higgins, V. J. (1993). Non-specific glycopeptide elicitors of *Cladosporium fulvum*: evidence for involvement of active oxygen species in elicitor-induced effects on tomato cell suspensions. *Physiological and Molecular Plant Pathology* **42**, 9–12.

Vera-Estrella, R., Barkia, B. J., Higgins, V. J. and Blumwald, E. (1994a). Plant defense response to fungal pathogens. Activation of host-plasma membrane H^+-ATPase by elicitor induced enzyme dephosphorylation. *Plant Physiology* **104**, 209–215.

Vera-Estrella, R., Higgins, V. J. and Blumwald, E. (1994a). Plant defense response to fungal pathogens. II. G-protein-mediated changes in host-plasma membrane redox reactions. *Plant Physiology* **106**, 97–102.

Walker, J. C. (1993). Receptor-like protein kinase genes of *Arabidopsis thaliana*. *The Plant Journal* **3**, 451–456.

Wevelsiep, L., Kogel, K. H. and Knogge, W. (1991). Purification and characterization of peptides from *Rhynchosporium secalis* inducing necrosis in barley. *Physiological and Molecular Plant Pathology* **39**, 471–482.

Wevelsiep, L., Rüpping, E. and Knogge, W. (1993). Stimulation of barley plasmalemma H^+-ATPase by phytotoxic peptides from the fungal pathogen *Rhynchosporium secalis*. *Plant Physiology* **101**, 297–301.

Whitham, S., Dinesh-Kumar, S. P., Choi, D., Hehl, R., Corr, C. and Baker, B. (1994). The product of the tobacco mosaic virus resistance gene *N*: similarity to Toll and the interleukin-1 receptor. *Cell* **78**, 1101–1115.

Wubben, J. P., Joosten, M. H. A. J., Van Kan, J. A. L. and De Wit, P. J. G. M. (1992). Subcellular localization of plant chitinases and 1,3-β-glucanases in *Cladosporium fulvum* (syn. *Fulvia fulva*)-infected tomato leaves. *Physiological and Molecular Plant Pathology* **41**, 23–32.

Wubben, J. P., Eijkelboom, C. A. and De Wit, P. J. G. M. (1994). Accumulation of pathogenesis-related proteins in epidermis of tomato leaves infected by *Cladosporium fulvum*. *Netherlands Journal of Plant Pathology* **99**, 231–239.

Wubben, J. P., Lawrence C. B. and De Wit, P. J. G. M. (1995). Differential induction of chitinase and 1,3-β-glucanase gene expression in tomato by *Cladosporium fulvum* and its race-specific elicitors. *Physiological and Molecular Plant Pathology* In press.

Phytoplasmas: Can Phylogeny Provide the Means to Understand Pathogenicity?

BRUCE C. KIRKPATRICK and CHRISTINE D. SMART

Department of Plant Pathology, University of California, Davis, California 95616, USA

Advances in Botanical Research Vol. 21
incorporating Advances in Plant Pathology
ISBN 0-12-005921-5

I. INTRODUCTION

During the past ten years the application of recombinant DNA technologies has greatly expanded our knowledge about the genetic diversity of plant pathogenic mycoplasma-like organisms (MLOs). During this same time there has been a veritable revolution in the field of bacterial phylogeny and taxonomy. The very recent elucidation of phylogenetic relationships among MLOs has suggested experimental approaches that could be undertaken to identify MLO genes that are important in plant pathogenesis and insect transmission. Although considerable work is still needed to develop genetic tools for identifying and manipulating MLO genes, such technologies have been successfully developed and used in phylogenetically similar organisms. This chapter presents an overview of MLO phylogeny and how this information may be used to further our knowledge about this important and unique group of plant pathogens.

A. MOLECULAR APPROACHES TO CHARACTERIZING MLOS

Since the discovery that phloem-limited, wall-less prokaryotes, and not viruses, were the causal agents of numerous graft transmissible "yellows diseases" (Doi et al., 1967; Ishiie et al., 1967), much effort has been expended trying to culture these pathogens in vitro (reviewed by Lee and Davis, 1986). Unfortunately, to date, all of these attempts have been unsuccessful. Although MLOs have been associated with more than 200 plant diseases worldwide (McCoy et al., 1989), the inability to culture them has been the single greatest impediment to characterizing their genetic diversity and relationships to other prokaryotes. The inability to culture these pathogens has also prevented the identification of MLO genes that are important in plant pathogenesis and transmission by insect vectors.

During the past ten years, several methodologies and technologies have been successfully used to isolate, detect, and characterize MLOs. Methods were developed to isolate MLO-enriched extracts from infected plants (Jiang and Chen, 1987; Kirkpatrick et al., 1987; Lee and Davis, 1988; Clark et al., 1989) and insect vectors (Kirkpatrick et al., 1987; Golino et al., 1989). These enriched extracts were used to produce MLO-specific polyclonal (Clark et al. 1983, 1989; Kirkpatrick and Garrott, 1984) and monoclonal (Lin and Chen, 1985; Jiang et al., 1989) antibodies that greatly facilitated the routine diagnosis of MLO-infected plants. Similarly, fragments of MLO chromosomal and extrachromosomal DNA were isolated and cloned from infected plants and insect vectors (reviewed by Kirkpatrick, 1992). These cloned fragments of MLO DNA were used as probes in DNA:DNA hybridization and restriction fragment length polymorphism (RFLP) analyses to compare the extent of genetic similarity between genetically similar MLOs (Lee and Davis, 1988; Lee *et al.*, 1989, 1991a, b, 1992a, b, 1993; Harrison *et al.*, 1991, 1992; Kuske *et al.*, 1991a; Ahrens *et al.*, 1993; Maurer *et al.*, 1993). This type of analysis provided considerable insight into the genetic diversity and relationships among some MLOs, especially those that had been transmitted to herbaceous plants such as periwinkle. However, this approach was not completely satisfactory because the production, characterization and use of numerous randomly cloned fragments of the MLO genome as probes in hybridization analyses was a time-consuming endeavor that gave only a partial picture of the diversity of these pathogens. Furthermore, this approach provided no information about relationships between MLOs and other culturable prokaryotes because none of the cloned fragments of MLO DNA hybridized with DNA from other prokaryotes (Kirkpatrick *et al.*, 1990). Clearly another approach was needed that would allow genetically diverse MLOs to be compared with each other and other prokaryotes, as well as permit characterization of the numerous low titer MLOs that infect woody plants, many of which have not been transmitted to periwinkle.

B.   PHYLOGENY AS A FRAMEWORK FOR PROKARYOTE TAXONOMY

During this same period of time a revolution was occurring in bacterial taxonomy. Powerful arguments were proposed by Woese (1987) and others that detailed analyses of evolutionarily conserved genes, especially the 16S ribosomal RNA, provided the means to identify phylogenetic relationships among widely divergent genera that share little or no homology among their genomes. Because rRNA is found in all cellular organisms and the function of these molecules must be conserved, the number of allowable alterations in the primary molecular sequence is limited. Thus they can serve as "molecular clocks" (Woese, 1987) whose analysis can provide insight into the evolutionary relationships of all cellular organisms. This approach, which has been greatly facilitated by the ability to directly amplify and sequence 16S rRNA genes with polymerase chain reaction (PCR) technologies, has had a tremendous impact on prokaryote phylogeny/taxonomy. A very large database of 16S rRNA sequences and widely available software for phylogenetic analyses now exist. The impact of molecular phylogenetic classification has had an even greater influence in understanding relationships amongst fastidious prokaryotes. This is especially so for members of the Class Mollicutes because their nutritional requirements are so complex. Here the utility of traditional bacterial classification parameters, which are largely based on defined substrate requirements or utilization, is greatly limited. However, one of the largest areas affected by molecular phylogeny has been the ability to classify and characterize nonculturable prokaryotes. Thus it is now possible to answer two fundamental questions about MLOs. (1) To what other prokaryotes are MLOs most closely related? and (2) What is the genetic diversity of the MLOs that are associated with over 200 plant diseases? During the past four years the analysis of MLO 16S rRNA and other evolutionarily conserved markers has provided considerable insight into these questions.

II. PHYLOGENY AND TAXONOMY OF MOLLICUTES AND MLOs

A. PHYLOGENETIC RELATIONSHIPS BASED ON 16S RIBOSOMAL RNA SEQUENCES

With the exception of *Thermoplasma*, most wall-less prokaryotes are members of the Class Mollicutes. To date seven genera are found within the Mollicutes: *Mycoplasma, Ureaplasma, Spiroplasma, Acholeplasma, Anaeroplasma, Asteroleplasma, Mesoplasma*, and *Entomoplasma* (Tully, 1989; Tully *et al.*, 1993). There are over 80 recognized *Mycoplasma* species, most of which are associated with vertebrate and insect hosts; however, a few species have been found as epiphytes on plant leaves and flowers (Rose *et al.*, 1990; Tully *et al.*, 1990). *Ureaplasma* s.p., *Anaeroplasma* s.p. and *Asteroleplasma* s.p. have only been found in association with animal hosts. Both culturable and non-culturable spiroplasmas are associated with various insect hosts (reviewed by Hackett and Clark, 1989) and the surfaces of leaves and flowers (Davis *et al.*, 1981). Three plant pathogenic spiroplasmas have been identified to date: *S. citri*, the causal agent of citrus stubborn (Fudl-Allah *et al.*, 1972; Saglio *et al.*, 1973) and horseradish brittleroot (Fletcher *et al.*, 1981) diseases; *S. kunkelii*, the causal agent of corn stunt disease (Chen and Liao, 1975; Williamson and Whitcomb, 1975); and *S. phoenicium*, a spiroplasma isolated from a diseased periwinkle plant growing in Iraq (Saillard *et al.*, 1987). *Acholeplasma* species have been recovered from animals, insects, soil compost, and plant surfaces (reviewed by Tully, 1989). The mollicute genera historically have been classified on the basis of morphology (spiroplasmas), metabolic (sterol requiring vs. non-requiring; aerobic vs. anaerobic), and membrane properties (digitoxin sensitivity). Species within a genus were differentiated primarily on the basis of antigenic properties and in some cases, DNA:DNA homologies. Although mollicutes lack a cell wall, early work by Neimark and London (1982) showed that they were most closely related to Gram-positive lactobacilli and originated by genome reductions from a walled ancestor.

Given the fact that most mollicutes lack any distinguishing morphological features and that serological and DNA hybridization analyses suggested that mollicutes were a diverse collection of wall-less organisms, additional criteria were needed to more accurately assess relationships among mollicute species and between mollicutes and other prokaryotes. Woese *et al.* (1985) analyzed 16S rRNA oligonucleotide signature sequences of several *Mycoplasma, Spiroplasma*, and *Acholeplasma* species and found that these organisms constituted a rapidly evolving group that was most closely related to several *Clostridium* species. However, the most comprehensive analysis of Mollicute phylogeny was performed by Weisberg *et al.* (1989) who cloned and sequenced nearly full-length 16S rRNA from approximately 50 mollicute species. They found that all of the mollicutes and their walled relatives formed a distinct clade within the low G + C Gram-positive bacteria. The 50 mollicute species could be organized into distinct groups or "clades": the *Mycoplasma hominis* clade; the *Mycoplasma pneumoniae* clade, which included *Ureaplasma*; the *Spiroplasma/ M. mycoides* clade; the *Acholeplasma/Anaeroplasma* clade; and a clade composed of a single species, *Asteroleplasma anaerobium* (Weisberg *et al.*, 1989). This analysis also revealed the apparent mis-classification of some *Mycoplasma* and *Acholeplasma* species. In order to address this problem two new genera, *Mesoplasma* and *Entomoplasma*, were proposed (Tully *et al.*, 1993).

The power of 16S rRNA analyses to resolve other problems of mollicute taxonomy was further reinforced in solving the mystery of the pommier organism (PPAV). PPAV is a wall-less prokaryote that was isolated from apple seeds taken from a tree with apple proliferation disease (Vignault *et al.*, 1980). PPAV, a non-helical mollicute which required sterols for growth, was serologically unrelated to any known *Mycoplasma* species. Because serological properties are the major determinants in classifying *Mycoplasma* species, it was initially assumed that PPAV would represent a new *Mycoplasma* species. In order to determine what mollicute this organism was most closely related to, its 16S rRNA gene was cloned and sequenced. Surprisingly, the PPAV sequence was 99.5% similar to *Mycoplasma iowae*, a mycoplasma pathogen of poultry! Apparently this particular isolate had lost or altered its primary antigenic determinant, thus rendering its serological classification meaningless (Grau *et al.*, 1991). This example, the lack of defined metabolic parameters for classifying mollicute species, and the comparatively large and growing database of mollicute 16S rRNA sequences, have made the determination of 16S rRNA sequences one of the most definitive approaches to classifying wall-less prokaryotes.

Because the power of molecular phylogenetic analyses was quickly recognized by mycoplasmologists, several research groups began cloning, sequencing, and characterizing 16S rRNA genes from non-culturable wall-less prokaryotes, especially plant pathogenic MLOs. Full-length sequences of three MLOs, Michigan aster yellows (O-MLO), California aster yellows (SAY), western X-disease (WX), were determined by 1992 and all of these pathogens were found to be unique members of the class Mollicutes (Lim and Sears, 1989; Kuske and Kirkpatrick, 1992).

Sequence analyses showed that the 16S rRNA genes of the O- and SAY-MLOs were 99.5% homologous. This result was not unexpected because these two MLOs both produce symptoms of phyllody and virescence in their host plants, and both are transmitted by the aster leafhopper *Macrosteles fascifrons*. In addition, cloned fragments of O-MLO chromosomal and plasmid

DNA cross-hybridized with SAY DNA (Sears *et al.*, 1989; Kuske *et al.*, 1991a, b). In contrast, the 16S rRNA sequence of WX was only 89% homologous to the O- and SAY-MLOs. This comparatively large value represents a significant divergence in sequence similarity and suggests that these two pathogens are quite different. This conclusion is also supported by biological properties such as plant symptomatology and vector specificity, as well as serological and DNA hybridization analyses (Lin and Chen, 1985; Kirkpatrick *et al.*, 1987; Lee *et al.*, 1992a). These results suggested that detailed analysis of just this one phylogenetic marker could provide considerable insight into characterizing the genetic diversity of MLOs.

   Phylogenetic analyses of these three MLOs clearly showed they were members of the Class Mollicutes; however, the MLOs were more closely related to the *Acholeplasma/Anaeroplasma* clade than to any of the clades containing true *Mycoplasma* species. In addition, these three MLOs were more closely related to each other than they were to any other culturable mollicute. Thus the trivial term "mycoplasma-like" had outlived its usefulness for at least two reasons. First, the designation "-like" traditionally implied a considerable degree of uncertainty concerning the affiliation of these pathogens. However, molecular phylogenetic data, which were rapidly becoming the hallmark criteria for classifying culturable mollicutes, clearly showed that the MLOs were mollicutes. Thus the term "-like" was clearly inappropriate. Secondly, MLOs were most closely related to *Acholeplasma* and *Anaeroplasma* and not true mycoplasmas. This conclusion has also been supported by MLO membrane properties (Lim *et al.*, 1992) and the detailed sequence of two other evolutionary markers, ribosomal protein genes (Lim and Sears, 1992; Gunderson *et al.*, 1994; Toth *et al.*, 1994) and 16/23S rRNA spacer regions (Kirkpatrick *et al.*, 1994). Additional MLO properties, such as genome size (Lim and Sears, 1991; Neimark and Kirkpatrick, 1993) and G+C content (Kollar and Seemuller, 1989) also showed that MLOs were most similar to the mollicutes. Finally, the unique ability to colonize plant phloem and be transmitted by certain phloem-feeding insects sets the MLOs apart biologically from culturable mollicute species. All of this evidence strongly supported the conclusion that the term MLO should be replaced by another, more accurate and formal designation. In 1992, a proposal was made to the Subcommittee on Taxonomy of Mollicutes of the International Organization of Mycoplasmology that the trivial term "Phytoplasma" be adopted in place of MLO (International Committee on Systematic Bacteriology, Subcommittee on the Taxonomy of Mollicutes, 1993). The Subcommittee was receptive to this proposal, but felt that because MLOs were associated with over 200 plant diseases, additional MLOs needed to be characterized in order to assess their potential heterogeneity.

   The 16S rRNA sequences of the O-MLO, SAY and WX were obtained by traditional approaches such as Southern blot analyses of healthy and MLO-infected plants to identify the fragment(s) containing the 16S rRNA gene, cloning these fragment(s) in *Escherichia coli*, screening the library to identify the clone of interest, preparing nested deletions of the cloned gene for sequence analysis and sequencing the deletions using standard dideoxy chain termination methods. Thus the determination of these first MLO 16S rRNA sequences was a fairly labor-intensive endeavor. Fortunately, the development and application of polymerase chain reaction (PCR) technologies greatly expedited the task of sequencing additional MLO 16S rRNA genes. Using oligonucleotide PCR primers obtained from sequences of 16S rRNA genes of culturable mollicutes, Deng and Hiruki (1991) were the first to PCR amplify nearly full-length MLO 16S rRNA genes from infected plants. Other researchers used different primers to amplify portions of the 16S rRNA genes from a large number of

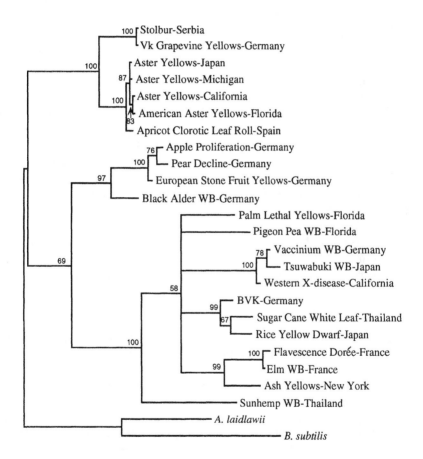

Fig. 1. Phylogram of 16S rDNA sequences for the major MLO groups based on a PAUP bootstrap analysis (Swofford, 1993). All MLOs analyzed to date form a monophyletic cluster which is most closely related to *Acholeplasma* species. However there is considerable evolutionary divergence among the major MLO groups. Numbers above the branches indicate bootstrap confidence values. *Acholeplasma laidlawii* and *Bacillus subtilis* were used as outgroups.

MLO isolates. The PCR-amplified products were then digested with various restriction endonucleases and the MLOs were classified into various RFLP groups (Ahrens and Seemuller, 1992; Lee *et al.*, 1993; Schneider *et al.*, 1993). Although this approach was useful for tentatively classifying an unknown MLO, detailed comparisons with other prokaryotes could only be accomplished by use of full-length 16S rRNA sequences.

During the past three years several laboratories have invested a tremendous amount of effort in characterizing the 16S rRNA sequences of numerous MLOS. Their recent publications on MLO phylogeny have provided even more convincing evidence that all of the MLOs examined to date form a coherent and unique clade within the Mollicutes (Gunderson *et al.*, 1994; Seemuller *et al.*, 1994). PCR technologies greatly facilitated this endeavor, either by providing amplified MLO 16S rRNA genes that were subsequently cloned into a vector and sequenced (Namba *et al.*, 1993; Seemuller *et al.*, 1994) or by directly sequencing the PCR-amplified product using cycle sequencing methods (Kirkpatrick *et al.*, 1992; Gunderson *et al.*, 1994). These efforts, combined with the previously characterized MLO sequences, have resulted in a comprehensive phylogeny based on the full-length sequence of approximately 30 MLO 16S rRNA genes. Fig. 1 shows a representative phylogenetic tree based on 16S rRNA sequences. Ten major MLO phylogenetic clades are shown: (1) stolbur; (2) aster yellows; (3) apple proliferation; (4) coconut lethal yellowing; (5) pigeon pea witches' broom; (6) X-disease; (7) rice yellow dwarf; (8) elm yellows; (9) ash yellows; and (10) sunhemp. In addition, Gunderson identified three additional clades: (11) loofah witches' broom; (12) clover proliferation; and (13) peanut witches' broom. Detailed analysis of these sequences identified several unique, MLO-specific oligonucleotide signatures that were not found in other culturable mollicutes, giving further support to the concept that MLOs represent a unique group of non-culturable mollicutes (Gunderson *et al.*, 1994).

B. PHYLOGENETIC RELATIONSHIPS BASED ON 16/23S SPACER REGION AND RIBOSOMAL PROTEIN GENE SEQUENCEs

Sequence analysis of two other evolutionary markers confirms the conclusions obtained by 16S rRNA phylogenetic analyses. Sequence analyses of ribosomal proteins rps3 and rpl2 showed that MLOs were most closely related to Acholeplasmas (Lim and Sears, 1992; Gunderson *et al.*, 1994; Toth *et al.*, 1994). In addition, it was found that MLOs, like Acholeplasmas, used UGG as a tryptophan codon, whereas Mycoplasmas and Spiroplasmas and Ureaplasmas use the stop codon UGA to code for tryptophan (Lim and Sears, 1992). Recent work in our laboratory has shown that the spacer region that separates the 16S from the 23S rRNA is also a reliable and convenient phylogenetic marker for both culturable (Smart *et al.*, 1994) and non-culturable (Kirkpatrick *et al.*,

1994) mollicutes. Sequence analysis of more than 60 MLO isolates showed that all of the 16/23S spacer regions (SR) contained one highly conserved tRNA^{Ile}, a structural organization that is unique among the mollicutes and other prokaryotes. Phylogenetic trees constructed from regions flanking the tRNA^{Ile} gene differentiated these MLOs into groups that were in complete agreement with the major MLO clades established by full-length 16S rRNA sequences (Kirkpatrick et al., 1994). Because the MLO SR is no more than 300 bp long it can be readily sequenced in both directions with only two sequencing primers. This fact has allowed us to characterize a larger number of MLO strains using SR sequences than have been analyzed by full-length 16S rRNA sequences (Kirkpatrick et al., 1994). Since the spacer region is probably under less evolutionary pressure than the rRNA structural genes, SR sequences could have greater utility for differentiating closely related MLOs than full-length 16S rRNA sequences.

In summary, a comparatively large database of MLO phylogenetic markers now exists. Phylogenetic relationships based on full-length 16S rRNA sequences, two ribosomal protein gene sequences, and the 16/23S spacer regions are all similar. These data support the conclusion that all the MLOs examined to date form a coherent, monophyletic group of organisms that is most closely related to members of the *Acholeplasma/Anaeroplasma* clade. Given this large amount of concordant experimental evidence, in 1994 the Working Team on MLOs of the International Organization of Mycoplasmology proposed to the Subcommittee on Taxonomy of the Mollicutes that the term "Phytoplasma" replace the trivial term MLO. The Taxonomy Subcommittee supported this proposal and recommended that the "Candidatus" status, as proposed by Murray and Schleifer (1994), be used when naming phytoplasma species. The major clades established by 16S rRNA sequence analysis will be tentatively classified as Phytoplasma species. Two recent publications have adopted this term (Gunderson et al., 1994; Sears and Kirkpatrick, 1994) and the Phytoplasma Working Team is preparing formal papers to authenticate this phylogenetically-based MLO taxonomy.

III. PROSPECTS FOR CHARACTERIZING AND EXPRESSING PHYTOPLASMA GENES

A. CULTURABLE HOSTS THAT MIGHT EXPRESS PHYTOPLASMA GENES

The inability to culture phytoplasmas *in vitro* has greatly impeded experimental manipulation of these pathogens. One approach to overcoming this difficulty would be to identify similar culturable organisms that could express phytoplasma genes. Phylogenetic analyses showed that several *Acholeplasma* species are most similar to phytoplasmas, which suggests acholeplasmas may be expression candidates. Acholeplasmas are easily cultured, they are morphologically similar to phytoplasmas, codon usage is similar in phytoplasmas and acholeplasmas (Lim and Sears, 1992) and transformation systems have been developed for acholeplasmas (see below). Another very attractive property of acholeplasmas is their ability to multiply within leafhoppers that are vectors of phytoplasmas. Whitcomb *et al.* (1973) injected *Acholeplasma laidlawii* and *A. granularum* into *Macrosteles fascifrons* and *Dalbulus elimatus* leafhopper vectors, and both organisms multiplied to high titers. These results suggest that a gene(s) that mediates phytoplasma insect transmission might be identified by transforming a library of phytoplasma DNA into an acholeplasma species that could multiply in a phytoplasma leafhopper vector. Transformants could be selected *in vitro* by use of antibiotics, injected into the vector and, following a suitable incubation period, the injected vectors could be fed on sachets containing a sucrose solution. If any of the acholeplasmas expressed phytoplasma genes that permitted their passage into the salivary gland cells, they could be recovered from the sachet solution. Although this specific example is quite speculative, the utility of acholeplasmas for expressing phytoplasma genes should be examined.

B. POTENTIAL TRANSPOSON AND PLASMID VECTORS FOR GENETICALLY MANIPULATING PHYTOPLASMAS

Given their small genome size and lack of a cell wall, members of the Mollicutes might seem to be ideal candidates for molecular genetic analyses. However, until recently it was not possible experimentally to introduce foreign DNA into these organisms (Bove *et al.*, 1984; Dybvig and Cassell, 1987; Dybvig, 1989; Mahairas *et al.*, 1990). The lack of efficient transformation or transfection procedures, vectors, and selectable markers were the primary constraints on molecular genetic characterization of the class *Mollicutes*. The first breakthrough in introducing foreign DNA into a mollicute was achieved by Sladek and Maniloff (1983) who successfully transfected *Acholeplasma laidlawii* with the replicative form (RF) DNA of acholeplasma virus L2. This transfection was achieved by treating the cells with polyethylene glycol (PEG) which is thought to alter the permeability of the cell membrane to exogenous DNA. Identical PEG-mediated transfection procedures were used to introduce a cloned spiroplasma virus (SpV4) into *Spiroplasma melliferum* (Pascarel-Devilder *et al.*, 1986). Native ssDNA and linearized RF of spiroplasma virus SVTS2 were also transfected into *S. citri* by a PEG method (McCammon and Davis, 1987). Cloned SVTS2 RF, if propagated in *E. coli* strain HB101 but not JM109, was also infectious either by PEG-mediated transformation or electroporation. The efficiency of transfection via PEG or electroporation was similar in these experiments. It was postulated that the loss of infectivity when the RF was propagated in JM109 was due to the presence of a Type I adenine modification system in JM109 (McCammon *et al.*, 1990). Electroporation was the only technique that allowed acholeplasma virus L3 to be transfected into

A. laidlawii (Lorenz *et al.*, 1988). It was postulated that the large size and linearity of this molecule prevented successful uptake by PEG-treated cells. However, electroporation-mediated transfection of a smaller acholeplasma virus (L1) was one-tenth less efficient than PEG-mediated transfection of L1 DNA. Thus it appears that the efficacy of both PEG and electroporation should be evaluated in attempts to develop a new transfection/transformation system for an uncharacterized mollicute/vector combination.

The first introduction of non-viral DNA into mollicutes was reported by Dybvig and Cassell (1987) who transformed PEG-treated *A. laidlawii* and *Mycoplasma pulmonis* cells with the Gram-positive streptococcal transposon Tn916, which encodes tetracycline resistance via the *tet*M gene. Southern blot analysis of several transformants showed that transposition of this element occurred at numerous places on the mollicute chromosome. It would appear that both staphylococcal and streptococcal transposons could be useful vectors for transposing wall-less mollicutes. The staphylococcal transposon Tn4001, which encodes gentamicin rather than tetracycline resistance, has been randomly inserted into the chromosome of PEG-treated *A. laidlawii* or *M. gallisepticum* cells (Dybvig and Cassell, 1987; Cao *et al.*, 1994). Once integrated into the chromosome the transconjugates were stable, even in the absence of gentamicin selection. Southern blot analysis of independently isolated transconjugates showed that Tn4001 inserted randomly throughout the chromosome (Cao *et al.*, 1994). Because of the random insertion of both Tn916 and Tn4001, it is quite possible that transposons could be used to inactivate and thus identify specific genes in these mollicutes.

Streptococcal plasmid vectors have also been studied for their ability to replicate in *Acholeplasma laidlawii*. The first report was that of Dybvig (1989), in which two streptococcal plasmids, pVA868 and pVA920, were transformed into *A. laidlawii*. Once in the acholeplasma, both of these large plasmids underwent deletion events to generate much smaller plasmids that contained the antibiotic markers and the origin of replication (ori). Although the transformation efficiency of the original streptococcal plasmids was low ($< 10^{-8}/$ CFU), the transformation efficiency of the smaller deletion plasmids was 100 times greater. *A. laidlawii* has also been transformed with the promiscuous lactic streptococcal vector pNZ18 (Sundstrom and Wieslander, 1990). The plasmid was maintained in *A. laidlawii* strain 8195 (a restriction-deficient strain) and was not modified or rearranged. In the absence of antibiotic selection pressure, however, the plasmid was lost. The transformation frequencies for pNZ18 reported by Sundstrom and Wieslander (1990) were higher ($4 \times 10^{-4}/$ CFU) than those reported for other streptococcal plasmids (Dybvig, 1989).

A novel approach for introducing selectable markers and foreign DNA into mollicutes was recently reported by Cao *et al.* (1994). Integrative plasmid vectors, which contained an *E. coli* replicon and the antibiotic resistance gene from either Tn916 or Tn4001, were ligated with restriction fragments derived from the *M. gallisepticum* chromosome. Recipient *M. gallisepticum* cells were transformed with the integrative plasmid vectors. Homologous recombination occurred between the chromosomal DNA in the vector and the recipient's chromosome and transconjugates were identified through antibiotic selection. These vectors allow one to establish whether a particular fragment of mollicute DNA contains genes of interest by altering it *in vitro* and evaluating the outcome of that mutation by introducing it back into the organism by homologous recombination. This transformation system has recently been used by Knudtson and Minion (1994) to identify upstream gene regulatory sequences in *Acholeplasma oculi*.

C. STRATEGIES FOR GENETICALLY MANIPULATING PHYTOPLASMAS

The preceding discussion clearly indicates that the number and variety of methods to introduce and manipulate genes of interest in wall-less prokaryotes has progressed greatly in recent years. The lack of a transformation system for phytoplasmas has slowed progress towards identifying and characterizing phytoplasma genes and gene products. Although the technical problems associated with a non-culturable mollicute greatly complicate the selection of individual transformants, it may still be possible. Because we now know that phytoplasmas are most closely related to the *Acholeplasmataceae*, initial transformation experiments could focus on vectors which have been successfully used to transform acholeplasmas.

Transposon Tn916 is an attractive candidate for phytoplasma transformation experiments because of the tetracycline resistance marker (*tet*M), which is functional in Gram-positive and Gram-negative bacteria as well as in mycoplasmas. Because phytoplasmas are insensitive to most antibiotics, but sensitive to tetracycline, it is crucial that a tetracycline resistance gene be present on any phytoplasma transformation vector. Many of the other plasmid and transposon elements that were described above could also be evaluated as potential phytoplasma vectors.

Although phytoplasmas cannot be cultured *in vitro*, viable phytoplasma extracts can be isolated from infected insects (Whitcomb *et al.*, 1966; Nasu *et al.*, 1974; Smith *et al.*, 1981). The cells can be concentrated by centrifugation, and transformation vectors introduced into the phytoplasma cell extract by use of PEG (Sladek and Maniloff, 1983) or electroporation (Gasparich *et al.*, 1993). Resulting putative transformants could then be injected into the appropriate leafhopper vector with fine glass needles (reviewed by Markham, 1982). Injected leafhoppers placed on indicator plants would transfer the phytoplasmas to these plants. Selection for transformants could then be accomplished by watering plants with dilute solutions of tetracycline. Because of the fastidious nature of phytoplasmas, it will be difficult to isolate a single transformant; however, relatively small numbers of transformed phytoplasmas could be selected by briefly feeding injected leafhoppers on test plants.

The discovery that some plant pathogenic phytoplasmas possess extrachromosomal DNA may represent one of the best opportunities to further expand molecular, genetic studies on these pathogens. Davis *et al.* (1988) were the first to identify and clone a fragment of extrachromosomal DNA from the maize bushy stunt phytoplasma (MBS). They reported that extrachromosomal DNA was present in Florida and Mexico strains of the MBS but absent in Texas strains. Most of these molecules were present in high enough titers to be visualized in agarose gels stained with ethidium bromide. Additional evidence supporting a comparatively high copy number of phytoplasma plasmids was provided when labeled cloned fragments of MBS extrachromosomal DNA were used as probes in DNA hybridization assays. Extrachromosomal probes provided more sensitive detection of the pathogen in plant and insect hosts than comparably sized, cloned probes derived from the MBS chromosome. Sears *et al.* (1989) found that one of the cloned fragments of phytoplasma DNA from an eastern strain of AY was extrachromosomal DNA. When labeled and used as a hybridization probe, it too provided superior pathogen detection. Extrachromosomal DNAs have also been identified in several members of the western X group of phytoplasmas including walnut witches' broom (Chen *et al.*, 1992) and vaccinium witches' broom (Schneider *et al.*, 1993). Because of the close phylogenetic relationship between these phytoplasmas and western X, it may be possible to transform western X with extrachromosomal DNA from walnut or vaccinium witches' broom. In this way, we may be able to determine if the extrachromosomal DNA has any affect on plant symptomatology and host range, or vector transmission.

Full-length phytoplasma plasmids have been cloned in *E. coli* vectors (Chen *et al.*, 1992; Kuske and Kirkpatrick, 1989; Schneider *et al.*, 1992). If these cloned molecules possess functional phytoplasma plasmid origins of replication they may be able to replicate in closely related phytoplasmas. It should also be possible to insert a tetracycline resistance gene which could function as a selectable marker in plants. The development of shuttle vectors that could be manipulated in *E. coli* and replicate in both phytoplasmas and *E. coli* would greatly facilitate molecular genetic analysis and manipulation of phytoplasmas.

IV. CANDIDATE MODEL SYSTEMS FOR IDENTIFYING PHYTOPLASMA PLANT PATHOGENICITY, HOST RANGE, AND INSECT TRANSMISSION GENES

Significant progress has been made in our understanding of molecular determinants of pathogenicity and plant hosts in some culturable, Gram-negative phytopathogenic prokaryotes. These accomplishments were made because these organisms were amenable to transformation with many *E. coli* vectors. In addition, most of these pathogenicity or host range determinants were identified by tagging and inactivating these genes with transposons. The ability to mutate an organism and characterize the genes responsible for a particular phenotype has been the hallmark approach used in molecular genetics.

Unfortunately, the inability to culture plant pathogenic phytoplasmas has made the identification, isolation, and propagation of strain mutants an extremely difficult task. The inability to mechanically inoculate phytoplasmas into plant phloem means that any potential mutant would still need to be transmissible by an insect vector. Fortunately, phytoplasma-enriched extracts can be prepared from infected insect vectors and these extracts retain their viability for several hours (reviewed by Markham, 1982). Thus it is possible to transform or otherwise manipulate the phytoplasmas in these extracts and then inject them back into appropriate insect vectors which can inoculate them into plant phloem. Unfortunately it is much more difficult to prepare infectious phytoplasma extracts from infected plants (Nasu et al., 1974). Some of the criteria for designing such experimental systems would include highly efficient insect acquisition from and transmission to plants, ease of raising host plants and insect vectors in the laboratory, and phytoplasma strains with clearly identifiable phenotypes in terms of plant symptomatology, host range, or insect transmission.

Examination of phytoplasma phylogenetic trees reveals several potential systems for isolating genes that determine plant host range (PHR genes), plant symptomatology (PS genes) and insect transmission (IT genes). Two central strategies can be envisioned for identifying and isolating these phytoplasma genes. The first strategy is to identify combinations of very closely related phytoplasmas, i.e. those in the same phylogenetic clade, but which differ considerably in PHR, PS or IT phenotypes. Recombinant molecules containing DNA from one strain would then be mobilized into the second strain and biological properties or antibiotic resistance would be used to enrich for the phenotype of interest. The second strategy involves the identification of very different phytoplasmas, i.e. those that lie in different phylogenetic clades and share little genetic homology, but which cause similar symptoms or are transmitted by the same insect vector. Genomic libraries of these organisms would be constructed in E. coli, screened with the heterologous DNA, and putative cloned phytoplasma genes further characterized to insure that the clones are unique and to identify functional genes that contain promoters and termination sequences. Potential PS or IT genes could then be cloned into phytoplasma transformation vectors, and transformed or electroporated into a phytoplasma which does not possess the desired phenotype. The insect, plant host, or antibiotic resistance is then used to select the desired phenotype. We will briefly describe four examples of such potential "model systems", if such a term is even appropriate for a non-culturable, phloem-limited, insect transmitted pathogen! Obviously the success of such an approach will depend on the development of suitable genetic vectors; a number of potential plasmid and transposon vectors have been discussed in the previous section.

A. THE ASTER YELLOWS/MAIZE BUSHY STUNT/*SPIROPLASMA KUNKELII*
SYSTEM

Phylogenetic analyses of aster yellows (AY) and maize bushy stunt (MBS)
phytoplasmas have shown that these pathogens are very closely related (Fig. 2).
This conclusion is also supported by DNA:DNA hybridization analyses with
cloned fragments of chromosomal (Kuske *et al.*, 1991a; N. Harrison, personal
communication) and extrachromosomal DNA (Kuske *et al.*, 1991b). However,
the AY and MBS phytoplasmas are very different in terms of their respective
insect vectors and plant hosts. Both MBS and *Spiroplasma kunkelii*, the causal
agent of corn stunt disease, are transmitted by the corn leafhopper *Dalbulus
maidis*, and certain other *Dalbulus* and *Baldulus* species (Madden and Nault,
1983). However, these maize leafhoppers do not transmit the AY phyto-

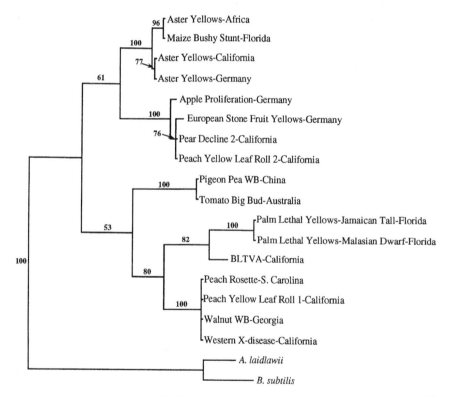

Fig. 2. Phylogram of 16S/23S spacer region sequences of 17 MLOs generated by
a PAUP bootstrap analysis (Swofford, 1993). Major phylogenetic groups delineated
above are similar to the major clades established by 16S rRNA sequences. Relationships
among the phytoplasmas shown here may provide the basis for identifying MLO plant
pathogenicity and host range genes. Numbers above the branches indicate the bootstrap
confidence values. *Acholeplasma laidlawii* and *Bacillus subtilis* were used as outgroups.

plasma. *D. maidis* preferentially feeds on maize and teosinte, and it will usually die if confined on any other plant for extended periods of time. Interestingly, the host plant specificity of *D. maidis* can be altered by feeding the insect on plants infected with the AY phytoplasma (Maramorosch, 1958; Purcell, 1988). Although the leafhopper does not transmit AY, leafhoppers fed on AY-infected plants can then be raised on a variety of plants in addition to corn. This "conditioning" of the corn leafhopper by AY may involve physiological alterations in the insect gut cells which may become locally infected with the AY pathogen. The AY phytoplasma is transmitted by several leafhopper species; however, it is transmitted very efficiently by the aster leafhopper, *Macrosteles fascifrons* (Kunkel, 1924). Although *M. fascifrons* can transmit the AY phytoplasma to a very wide variety of plants (Kunkel, 1926), the AY phytoplasma was not transmitted to corn by infectious *M. fascifrons* (Kuske *et al.*, 1991b). Although MBS and *S. kunkelii* are distantly related in terms of phylogeny and genomic homology, they both occupy the same biological niche. Both have A:T rich genomes and, therefore, they may possess similar IT and PHR genes.

Fig. 3 outlines a strategy for identifying IT and PHR genes in the MBS phytoplasma. One recent accomplishment that will greatly facilitate the preparation of phytoplasma genomic libraries was the development of techniques to isolate full-length phytoplasma chromosomes free from contaminating plant DNA by use of pulsed field gel electrophoresis (PFGE) (Neimark and Kirkpatrick, 1993). A library of PFGE-purified MBS DNA would be constructed in *E. coli* by using a suitable plasmid or lambda cloning vector. The library would then be screened with ^{32}P-labeled *S. kunkelii* DNA, which can be readily obtained in a pure form because this pathogen is culturable. If necessary, cross-hybridizing MBS clones could be further screened to eliminate "housekeeping" genes that would be found in other non-related phytoplasmas, such as western X. MBS clones that survived this screening process would be further characterized by restriction mapping and Southern blot analysis to identify unique cloned fragments of cross-hybridizing MBS DNA. These fragments would be sequenced to identify functional genes, i.e. those which contain putative promoter and termination sequences. Functional cross-hybridizing MBS genes would then be cloned into AY transformation vectors and transformed into infectious extracts of the AY phytoplasma. The transformed extracts would be injected into *Macrosteles fascifrons* or *D. maidis* that had been "preconditioned" to survive on aster plants (see above). The injected *D. maidis* leafhoppers would be incapable of transmitting AY into the aster plant unless the AY phytoplasma had been transformed with MBS gene(s) that determine IT specificity. The injected *M. fascifrons* would be capable of transmitting AY; however, only those AY phytoplasmas that acquired genes which permitted corn phloem colonization would infect corn. If multiple MBS clones were initially obtained then these MBS genes could be combined and introduced into AY in various combinations. Although there

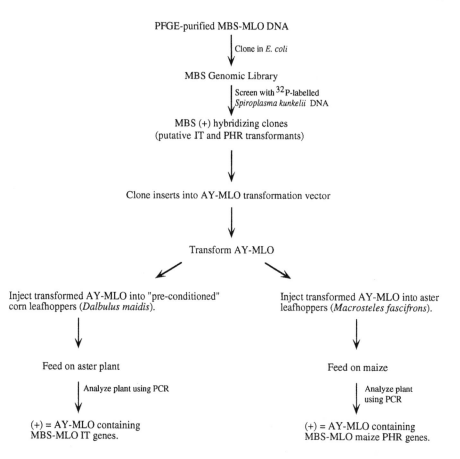

Fig. 3. Strategy for identifying insect transmission (IT) and plant host range (PHR) genes in maize bushy stunt (MBS)-MLO.

are numerous factors that could influence the outcome of these experiments, the overall approach seems feasible.

B. THE ASTER YELLOWS/BEET LEAFHOPPER TRANSMITTED VIRESCENCE AGENT/*SPIROPLASMA CITRI* SYSTEM

Spiroplasma citri and the beet leafhopper transmitted virescence agent (BLTVA) are transmitted by the beet leafhopper *Circulifer tenellus* (Golino *et al.*, 1987). Both pathogens have a wide plant host range; however, BLTVA causes phyllody and virescence (Golino *et al.*, 1989) but *S. citri* does not (Calavan and Oldfield, 1979). The AY phytoplasma infects and causes virescence and phyllody in most of the same hosts as BLTVA. However, the AY phytoplasma is not transmitted by *C. tennelus* and the BLTVA phytoplasma is not transmitted by *M. fascifrons*. Furthermore, the AY and BLTVA phytoplasmas, and *S. citri* are phylogenetically (Fig. 2) and genetically dissimilar (Lee *et al.*, 1990; Kuske and Kirkpatrick, 1991a, b). It is conceivable that these genetic differences and biological similarities could be exploited to identify virescence and phyllody genes as well as IT genes of the BLTVA phytoplasma by the strategy outlined in Fig. 4.

A library of PFGE-purified BLTVA DNA would be constructed in *E. coli* with suitable plasmid or phage vectors. The library would be screened with ^{32}P-labeled AY and *S. citri* DNA at low stringencies. Depending on the number of cross-hybridizing clones obtained, their numbers could be further

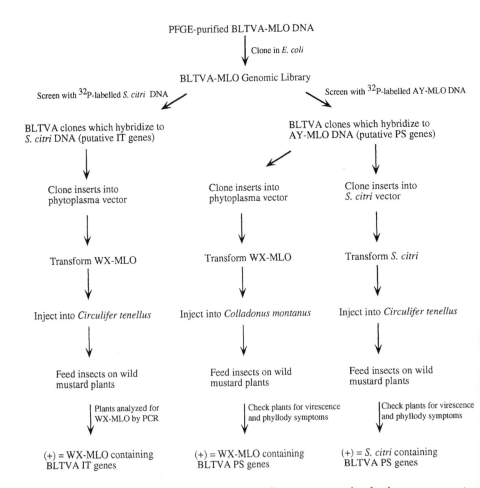

Fig. 4. Strategy for identifying beet leafhopper transmitted virescense agent (BLTVA)-MLO insect transmission (IT) and plant symptomology (PS) genes.

reduced by screening with DNA from a phytoplasma, such as western X-disease (WX) which does not cause virescence and phyllody nor is it transmitted by *C. tenellus*. Clones that survive the screening process would then be further characterized as previously described for the MBS system. Functional candidate PS genes, i.e. genes responsible for the production of virescence and phyllody, or *C. tenellus* IT genes, could then be cloned into suitable phytoplasma and *S. citri* transformation vectors. Phytoplasma vectors could be transformed into a phytoplasma strain that does not cause virescence and phyllody, such as WX, whereas the *S. citri* vectors would be transformed into their host. *C. tennellus* injected with BLTVA-transformed *S. citri* cells would then be fed on a preferred host such as wild mustard which shows conspicious BLTVA symptoms (Golino *et al.*, 1989). WX phytoplasmas transformed with putative BLTVA IT and PS genes would be injected into both *C. tenellus* (a non-vector of WX) and *Colladonus montanus* (an efficient vector of X). Injected *C. tenellus* would be fed on a WX compatible host such as wild mustard and the plants analyzed for the presence of WX phytoplasma that acquired BLTVA IT genes. Injected *C. montanus* could also be fed on mustard which would show symptoms of virescence and phyllody if the transformed WX phytoplasma had acquired BLTVA PS genes.

C. THE WALNUT WITCHES' BROOM/X-DISEASE SYSTEM

As shown in Fig. 2, the X-disease and walnut witches' broom (WWB) phytoplasma are phylogenetically very similar. Their close relationship is also supported by serological and DNA:DNA hybridization analyses (Kirkpatrick *et al.*, 1990; Chen *et al.*, 1992). Interestingly, the WWB phytoplasma possesses at least two plasmids whereas no plasmids have been found in the WX phytoplasma (Kirkpatrick, unpublished results). These two phytoplasmas also differ in their ability to infect plant hosts; WX cannot infect black walnut, whereas walnut and pecan are the only known hosts of the WWB phytoplasma. The close genetic similarity of these two phytoplasmas suggests that walnut plant host range (PHR) genes could be expressed in the WX phytoplasma. A comparatively straightforward attempt to identify walnut PHR genes would be to prepare a library of WWB DNA in a suitable WX vector, transform infectious WX phytoplasma leafhopper extracts, inject the transformed extracts into an efficient leafhopper vector of WX, such as *C. montanus*, and feed the injected leafhoppers on black walnut seedlings. The walnut host would then select for any WX phytoplasmas that carried WWB PHR genes.

D. THE X-DISEASE/PEACH YELLOW LEAF ROLL SYSTEM

Peach yellow leaf roll (PYLR) disease can be caused by two phylogenetically and genetically unrelated phytoplasmas, shown as PYLR-1 and 2 in Fig. 2.

Symptoms of this disease are very conspicuous: the midrib of the leaf rolls downwards while the leaf lamina cups upwards, then leaves become very chlorotic and the midrib and lateral veins take on a conspicuous swollen appearance. PYLR-1 is transmitted by leafhoppers such as *Colladonus montanus*, whereas PYLR-2 is transmitted by pear psylla and it cannot be transmitted by *C. montanus* (Purcell and Suslow, 1982). Phytoplasma genes which cause the conspicuous PYLR leaf symptoms might be identified by screening a library of PYLR-1 with ^{32}P-labeled PYLR-2 DNA. Cross-hybridizing clones could be evaluated as previously described and candidate PYLR plant symptomatology genes transformed into infectious extracts of the WX phytoplasma. Transformed WX phytoplasmas could be injected into *C. montanus* which would be fed on peach seedlings. If the phytoplasma transformation vector possessed a tetracycline resistance gene, transformed WX could be selected by treating the peach seedling with foliar applications of tetracycline.

V. SUMMARY

The previous examples are certainly speculative approaches to identifying phytoplasma genes that mediate plant pathogenesis and insect transmission. However, the rapid progress that has been made in understanding the culturable prokaryotes to which the phytoplasmas are most closely related has rendered such approaches at least possible. The advances that are being made in developing transformation systems for closely related *Acholeplasma* species and the discovery of extrachromosomal elements in phytoplasmas offer hope that such experiments may be conducted before long. The development of methods to mechanically inoculate phytoplasmas into plant phloem, and of course to culture these pathogens *in vitro*, would obviously greatly facilitate the identification of phytoplasma pathogenesis and transmission genes.

REFERENCES

Ahrens, U. and Seemuller, E. (1992). Detection of DNA of plant pathogenic mycoplasmalike organisms by a polymerase chain reaction that amplifies a sequence of the 16S RNA gene. *Phytopathology* **82**, 828–832.

Ahrens, U., Lorenz, K.-H. and Seemuller, E. (1993). Genetic diversity among mycoplasmalike organisms associated with stone fruit diseases. *Molecular Plant-Microbe Interactions* **6**, 686–691.

Bove, J. M., Candresse, T., Mouches, C., Renaudin, J. and Saillard, C. (1984). Spiroplasmas and the transfer of genetic material by transformation and transfection. *Israeli Journal of Medical Science* **20**, 836–839.

Calavan, E. C. and Oldfield, G. N. (1979). Symptomatology of spiroplasmal plant disease. *In* "The Mycoplasmas", Vol. 3 (R. F. Whitcomb and J. G. Tully, eds), pp. 37–64. Academic Press, New York.

Cao, J., Kapke, P. A. and Minion, F. C. (1994). Transformation of *Mycoplasma galli-*

septicum with Tn916, Tn4001, and integrative plasmid vectors. *Journal of Bacteriology* **176**, 4459–4462.

Chen, T. A. and Liao, C. H. (1975). Corn stunt spiroplasma: isolation, cultivation and proof of pathogenicity. *Science* **188**, 1015–1017.

Chen, J., Chang, C. J., Jarret, R. and Gawel, N. (1992). Isolation and cloning of DNA fragments from a mycoplasmalike organism associated with walnut witches'-broom disease. *Phytopathology* **82**, 306–309.

Clark, M. F., Barbara, D. J. and Davies, D. L. (1983). Production and characteristics of antisera to *Spiroplasma citri* and clover phyllody associated antigens derived from plants. *Annals of Applied Biology* **103**, 251–259.

Clark, M. F., Morten, A. and Buss, S. L. (1989). Preparation of mycoplasma immunogens from plants and a comparison of polyclonal and monoclonal antibodies made against primula yellows-MLO-associated antigens. *Annals of Applied Biology* **114**, 111–124.

Davis, M. J., Tsai, J. H., Cox, R. L., McDaniel, L. L. and Harrison, N. A. (1988). Cloning of chromosomal and extrachromosomal DNA of the mycoplasma-like organisms that causes maize bushy stunt disease. *Molecular Plant-Microbe Interactions* **1**, 295–302.

Davis, R. E., Lee, I.-M. and Worley, J. F. (1981). *Spiroplasma floricola*, a new species isolated from the surface of flowers of the tulip tree (*Liriodendron tulipifera*). *International Journal of Systematic Bacteriology* **31**, 456–464.

Deng, S. and Hiruki, C. (1991). Amplification of 16S rRNA genes from culturable and nonculturable mollicutes. *Journal of Microbiological Methods* **14**, 53–61.

Doi, T., Teranaka, M., York, K. and Asuyama, H. (1967). Mycoplasma or PLT group organisms found in the phloem of plants infected with mulberry dwarf, potato witches' broom, aster yellows and Paulowinia witches' broom. *Annals of the Phytopathological Society of Japan* **33**, 256–266.

Dybvig, K. (1989). Transformation of *Acholeplasma laidlawii* with streptococcal plasmids pVA868 and pVA920. *Plasmid* **21**, 155–160.

Dybvig, K. and Cassell, G. H. (1987). Transposition of Gram positive transposon Tn916 in *Acholeplasma laidlawii* and *Mycoplasma pulmonis*. *Science* **235**, 1392–1394.

Fletcher, J., Schultz, G. A., Davis, R. E., Eastman, C. E. and Goodman, R. M. (1981). Brittle root disease of horseradish: evidence for an etiological roll of *Spiroplasma citri*. *Phytopathology* **71**, 1073–1080.

Fudal-Allah, A. F. S. A., Calavan, E. C. and Igwebge, E. C. K. (1972). Culture of a mycoplasma-like organism associated with stubborn disease of citrus. *Phytopathology* **62**, 729–731.

Gasparich, G. E., Hackett, K. J., Stamburski, C., Renaudin, J. and Bove, J. M. (1993). Optimization of methods for transfecting *Spiroplasma citri* strain R8A2 HP with the *Spiroplasma* virus SpV1 replicative form. *Plasmid* **29**, 193–205.

Golino, D. A., Oldfield, G. N. and Gumpf, D. J. (1987). Transmission characteristics of the beet leafhopper transmitted virescence agent. *Phytopathology* **77**, 954–957.

Golino, D. A., Kirkpatrick, B. C. and Fisher, G. A. (1989). The production of a polyclonal antisera to the beet leafhopper transmitted virescence agent. *Phytopathology* **79**, 1138.

Golino, D. A., Oldfield, G. N. and Gumpf, D. J. (1989). Experimental hosts of the beet leafhopper transmitted virescence agent. *Plant Disease* **73**, 850–854.

Grau, O., Laigret, F., Carle, P., Tully, J. G., Rose, D. L., Bove, J. M. (1991). Identification of a plant-derived mollicute as a strain of an avian pathogen, *Mycoplasma iowae*, and its implications for molicute taxonomy. *International Journal of Systematic Bacteriology* **41**, 473–478.

Gunderson, D. E., Lee, I.-M., Rehner, S. A., Davis, R. E. and Kingsbury, D. T. (1994). Phylogeny of mycoplasmalike organisms (phytoplasmas): a basis for their classification. *Journal of Bacteriology* **176**, 5244–5254.

Hackett, K. J. and Clark, T. B. (1989). Ecology of spiroplasmas. *In* "The Mycoplasmas", Vol. 5 (R. F. Whitcomb and J. G. Tully, eds), pp. 113–200. Academic Press, New York.

Harrison, N. A., Tsai, J. H., Bourne, C. M. and Richardson, P. A. (1991). Molecular cloning and detection of chromosomal and extra-chromosomal DNA of mycoplasmalike organisms associated with witches'-broom disease of pigeon pea in Florida. *Molecular Plant-Microbe Interactions* **4**, 300–307.

Harrison, N. A., Bourne, C. M., Cox, R. L., Tsai, J. H. and Richardson, P. A. (1992). DNA probes for detection of mycoplasma like organisms associated with lethal yellowing disease of palms in Florida. *Phytogathology* **82**, 216–224.

International Committee on Systematic Bacteriology, Subcommittee on the Taxonomy of *Mollicutes* (1993). Minutes of the interim meetings, 1 and 2 August 1992, Ames, Iowa. *International Journal of Systematic Bacteriology* **43**, 394–397.

Ishiie, T., Doi, T., York, K. and Asuyama, H. (1967). Suppressive effects of antibiotics of the tetracycline group on the symptom development in mulberry dwarf disease. *Annals of the Phytopathological Society of Japan* **33**, 267–275.

Jiang, Y. P. and Chen, T. (1987). Purification of mycoplasma-like organisms from lettuce with aster yellows disease. *Phytopathology* **77**, 949–953.

Jiang, Y. P., Lei, J. D. and Chen, T. A. (1988). Purification of aster yellows agent from diseased lettuce using affinity chromatography. *Phytopathology* **78**, 828–831.

Jiang, Y. P., Chen, T. A. and Chiykowski, L. N. (1989). Production of monoclonal antibodies to peach eastern X-disease agent and their use in disease detection. *Canadian Journal of Plant Pathology* **11**, 325–331.

Kirkpatrick, B. C. (1992). Mycoplasma-like organisms: plant and invertebrate pathogens. *In* "The Prokaryotes", second edition (A. Balows, H. G. Truper, M. Dworkin, W. Harder, and K. H. Schleifer, eds), pp. 4050–4067. Springer-Verlag, New York.

Kirkpatrick, B. C. and Garrott, D. G. (1984). Detection of western X-disease mycoplasma-like organisms by enzyme-linked immunosorbent assay. *Phytopathology* **74**, 825.

Kirkpatrick, B. C., Stenger, D. C., Morris, T. J. and Purcell, A. H. (1987). Cloning and detection of DNA from a nonculturable plant pathogenic mycoplasma-like organism. *Science* **238**, 197–200.

Kirkpatrick, B. C., Fisher, G. A., Fraser, J. D. and Purcell, A. H. (1990). Epidemiological and phylogenetic studies on western X-disease mycoplasma-like organisms. *In* "Recent Advances in Mycoplasmology" (G. Stanek, G. H. Casell, J. G. Tully, and R. F. Whitcomb, eds), pp. 288–297. *International Journal of Medical Microbiology*, Series A, Supplement 20.

Kirkpatrick, B. C., Gao, J. and Harrison, N. (1992). Phylogenetic relationships of 15 MLOs established by PCR sequencing of variable regions within the 16S ribosomal RNA gene. *Phytopathology* **82**, 1083.

Kirkpatrick, B. C., Smart, C. D., Gardner, S. L., Gao, J.-L., Ahrens, U., Maurer, R., Schneider, B., Lorenz, K.-H., Seemuller, E., Harrison, N. A., Namba, S. and Daire, X. (1994). Phylogenetic relationships of plant pathogenic MLOs established by 16/23S rDNA spacer sequences. *IOM Letters* **3**, 228–229.

Knudtson, K. L. and Minion, F. C. (1994). Use of *lac* gene fusions in the analysis of *Acholeplasma* upstream gene regulatory sequences. *Journal of Bacteriology* **176**, 2763–2766.

Kollar, A. and Seemuller, E. (1989). Base composition of the DNA of mycoplasma-like organisms associated with various plant diseases. *Journal of Phytopathology* **127**, 177–186.

Kunkel, L. O. (1924). Insect transmission of aster yellows agent. *Phytopathology* **14**, 54.

Kunkel, L. O. (1926). Studies on aster-yellows disease. *American Journal of Botany* **13**, 646–705.

Kuske, C. R. and Kirkpatrick, B. C. (1989). Identification and partial characterization of plasmids from the western aster yellows mycoplasma-like organism. *Journal of Bacteriology* **172**, 1628–1633.

Kuske, C. R. and Kirkpatrick, B. C. (1992). Phylogenetic relationships between the western aster yellows mycoplasmalike organisms and other prokaryotes established by 16S rRNA gene sequence. *International Journal of Systematic Bacteriology* **42**, 226–233.

Kuske, C. R., Kirkpatrick, B. C. and Seemuller, E. (1991a). Differentiation of virescence MLOs using aster yellows mycoplasma-like chromosomal DNA probes and restriction fragment length polymorphism analysis. *Journal of General Microbiology* **137**, 153–159.

Kuske, C. R., Kirkpatrick, B. C., Davis, M. J. and Seemuller, E. (1991b). DNA hybridization between western aster yellows mycoplasma-like organism plasmids and extrachromosomal DNA from other plant pathogenic mycoplasma-like organisms. *Molecular Plant–Microbe Interactions* **4**, 75–80.

Lee, I.-M. and Davis, R. E. (1986). Prospects for *in vitro* culture of plant pathogenic mycoplasma-like organisms. *Annual Review of Phytopathology* **24**, 339–354.

Lee, I.-M. and Davis, R. E. (1988). Detection and investigation of genetic relatedness among aster yellows and other mycoplasma-like organisms by using cloned DNA and RNA probes. *Molecular Plant–Microbe Interactions* **1**, 303–311.

Lee, I.-M., Davis, R. E., Chen, T. A., Chiykowski, L. N., Fletcher, J. and Hiruki, C. (1989). Nucleic acid hybridization distinguishes MLO strains transmitted by *Macrosteles* spp., vectors of aster yellows agent. *Phytopathology* **79**, 1137.

Lee, I.-M., Davis, R. E. and Hiruki, C. (1990). Beet leafhopper transmitted virescence agent and clover proliferation mycoplasma-like organisms (MLOs): two new distinct strain types. *Phytopathology* **80**, 958.

Lee, I.-M., Davis, R. E. and Hiruki, C. (1991a). Genetic relatedness among clover proliferation mycoplasmalike organisms (MLOS) and other MLOs investigated by nucleic acid hybridization and restriction fragment length polymorphism analyses. *Applied and Environmental Microbiology* **57**, 3565–3569.

Lee, I.-M., Davis, R. E., Chen, T. A., Chiykowski, L. N., Fletcher, J. and Hiruki, C. (1991b). Classification of MLOs in the aster yellows MLO strain cluster on the basis of RFLP analyses. *Phytopathology* **81**, 1169.

Lee, I.-M., Gundersen, D. E., Davis, R. E. and Chiykowski, L. N. (1992a). Identification and analysis of a genomic strain cluster of mycoplasmalike organisms, associated with Canadian peach (Eastern) X disease, Western X disease and clover yellow edge. *Journal of Bacteriology* **174**, 6694–6698.

Lee, I.-M., Davis, R. E., Chen, T. A., Chiykowski, L. N., Fletcher, J., Hiruki, C. and Schaff, D. A. (1992b). A genotype-based system for identification and classification of mycoplasmalike organisms (MLOs) in the aster yellows MLO strain cluster. *Phytopathology* **82**, 977–986.

Lee, I.-M., Hammond, R. W., Davis, R. E. and Gundersen, D. E. (1993). Universal amplification and analysis of pathogen 16S rDNA for classification and identification of mycoplasmalike organisms. *Phytopathology* **83**, 834–842.

Lim, P.-O. and Sears, B. B. (1989). 16S rRNA sequence indicates plant pathogenic mycoplasma-like organisms are evolutionarily distant from animal mycoplasmas. *Journal of Bacteriology* **171**, 1233–1235.

Lim, P.-O. and Sears, B. B. (1991). The genome size of a plant–pathogenic myco-plasmalike organism resembles those of animal mycoplasmas. *Journal of Bacteriology* **173**, 2128–2130.

Lim, P.-O. and Sears, B. B. (1992). Evolutionary relationships of a plant–pathogenic mycoplasmalike organism and *Acholeplasma laidlawii* deduced from two ribo-somal protein gene sequences. *Journal of Bacteriology* **174**, 2606–2611.

Lim, P.-O., Sears, B. B. and Klomparens, K. L. (1992). Membrane properties of a plant-pathogenic mycoplasmalike organism. *Journal of Bacteriology* **174**, 682–686.

Lin, C. P. and Chen, T. A. (1985). Monoclonal antibodies against the aster yellows agent. *Science* **227**, 1233–1235.

Lorenz, A., Just, W., da Silva Cardoso, M. and Klotz, G. (1988). Electroporation-mediated transfection of *Acholeplasma laidlawii* with mycoplasma virus L1 and L3 DNA. *Journal of Virology* **62**, 3050–3052.

Madden, L. V. and Nault, L. R. (1983). Differential pathogenicity of corn stunting mollicutes to leafhopper vectors in *Dalbulus* and *Baldulus* species. *Phytopatho-logy* **73**, 1608–1614.

Mahairas, G. G., Jean, C. and Minion, F. C. (1990). Development of a cloning system for *Mycoplasma pulmonis*. *Gene* **93**, 61–65.

Maramorosch, K. (1958). Beneficial effect of virus-diseased plants on nonvector insects. *Tijdschr Plantenziek* **64**, 383–391.

Markham, P. G. (1982). Insect vectors. *In* "Plant and Insect Mycoplasma Techniques" (M. J. Daniels and P. G. Markham eds), pp. 369. Croom Helm, London.

Maurer, R., Seemuller, E. and Sinclair, W. A. (1993). Genetic relatedness of myco-plasmalike organisms affecting elm, alder, and ash in Europe and North America. *Phytopathology* **83**, 971–976.

McCammon, S. L. and Davis, R. E. (1987). Transfection of *Spiroplasma citri* with DNA of a new-rod-shaped spiroplasma virus. *In* "Plant Pathogenic Bacteria" (E. L. Civerolo, A. Collmer, R. E. Davis, and A. G. Gillaspie, eds), pp. 458–464. Martinus Nijhoff, Dordrecht, The Netherlands.

McCammon, S. L., Dally, E. L. and Davis, R. E. (1990). Electroporation and DNA methylation effects on the transfection of *Spiroplasma*. *In* "Recent Advances in Mycoplasmology, Proceedings of the 7th Congress of the International Organization for Mycoplasmology" (G. Stanek, G. H. Cassell, J. G. Tully and R. F. Whitcomb, eds), pp. 60–66. Gustav Fischer Verlag, Stuttgart, Germany.

McCoy, R. E., Caudwell, A., Chang, C. J., Chen, T. A., Chiykowski, L. N., Cousin, M. T., Dale, J. L., de Leeuw, G. T. N., Golino, D. A., Hackett, K. J., Kirkpatrick, B. C., Marwitz, R., Petzold, H., Sinha, R. C., Sugiura, M., Whit-comb, R. F., Yang, I. L., Zhu, B. M. and Seemuller, E. (1989). Plant diseases associated with mycoplasma-like organisms. *In* "The Mycoplasmas", Vol. V (R. F. Whitcomb and J. G. Tully, eds), pp. 546–640. Academic Press, New York.

Murray, R. G. E. and Schleifer, K. H. (1994). Taxonomic notes; a proposal for recor-ding the properties of putative taxs of procaryotes. *International Journal of Systematic Bacteriology* **44**, 174–176.

Namba, S., Oyaizu, H., Kato, S., Iwanami, S. and Tsuchizaki, T. (1993). Phylogenetic diversity of phytopathogenic mycoplasmalike organisms. *International Journal of Systematic Bacteriology* **43**, 461–467.

Nasu, S., Jensen, D. D. and Richardson, J. (1974). Extraction of western X-disease mycoplasma-like organisms from leafhoppers and celery infected with peach western X-disease. *Applied Entomology and Zoology* **9**, 53–57.

Neimark, H. and Kirkpatrick, B. C. (1993). Isolation and characterization of full-length chromosomes from non-culturable plant-pathogenic mycoplasma-like organisms. *Molecular Microbiology* **7**, 21–28.

Neimark, H. and London, J. (1982). Origins of the mycoplasmas: sterol-nonrequiring mycoplasmas evolved from streptococci. *Journal of Bacteriology* **150**, 1259–1265.

Pascarel-Devilder, M-C., Renaudin, J. and Bove, J. M. (1986). The spiroplasma virus 4 replicative form cloned in *Escherichia coli* transfects spiroplasmas. *Virology* **151**, 390–393.

Purcell, A. H. (1988). Increased survival of *Dalbulus maidis*, a specialist on maize, on non-host plant infected with mollicute pathogens. *Entomology Experimental and Applied* **46**, 187–196.

Purcell, A. H. and Suslow, K. G. (1982). Surveys of leafhoppers (Homoptera: Cicadellidae) and pear psylla (Homoptera: Psyllidae) in pear and peach orchards relative to the spread of peach yellow leaf roll disease. *Journal of Economic Entomology* **77**, 1489–1494.

Rose, D. L., Kocka, J. P., Somerson, N., Tully, J. G., Whitcomb, R. F., Carle, P., Bove, J. M., Colflesh, D. E. and Williamson, D. L. (1990). *Mycoplasma lactucae*, sp. nov., a sterol-requiring mollicute from a plant surface. *International Journal of Systematic Bacteriology* **40**, 138–142.

Saglio, P., L'Hospital, M., Lafleche, D., Dupont, G., Bove, J. M., Tully, J. G. and Freundt, E. A. (1973). *Spiroplasma citri* gen. and sp. n.: a mycoplasma-like organism associated with stubborn disease of citrus. *International Journal of Systematic Bacteriology* **23**, 191–204.

Saillard, C., Vignault, J. C., Bove, J. M., Raie, A., Tully, J. G., Williamson, D. L., Fos, A., Garnier, M., Gadeau, A., Carle, P. and Whitcomb, R. F. (1987). *Spiroplasma phoeniceum* sp. nov., a new plant-pathogenic species from Syria. *International Journal of Systematic Bacteriology* **37**, 106–115.

Schneider, B., Maurer, R., Saillard, C., Kirkpatrick B. C. and Seemuller, E. (1992). Occurrence and relatedness of extrachromosomal DNAs in plant pathogenic mycoplasmalike organisms. *Molecular Plant-Microbe Interactons* **5**, 489–495.

Schneider, B., Ahrens, U., Kirkpatrick B. C. and Seemuller, E. (1993). Classification of plant-pathogenic mycoplasma-like organisms using restriction-site analysis of PCR-amplified 16S rDNA. *Journal of General Microbiology* **139**, 519–527.

Sears, B. B. and Kirkpatrick, B. C. (1994). Unveiling the evolutionary relationships of plant-pathogenic mycoplasmalike organisms. *ASM News* **60**, 307–312.

Sears, B. B., Lim, P-O., Holland, N., Kirkpatrick, B. C. and Klomparens, K. L. (1989). Isolation and characterization of DNA from a mycoplasma-like organism: *Molecular Plant-Microbe Interactions* **2**, 175–180.

Seemuller, E., Schneider, B., Maurer, R., Ahrens, U., Daire, X., Kison, H., Lorenz, K.-H., Firrao, G., Avinent, L., Sears, B. B. and Stackebrandt, E. (1994). Phylogenetic classification of phytopathogenic mollicutes by sequence analysis of 16S ribosomal DNA. *International Journal of Systematic Bacteriology* **44**, 440–446.

Sladek, T. L. and Maniloff, J. (1983). Polyethylene glycol-dependent transfection of *Acholeplasma laidlawii* with mycoplasma virus L2 DNA. *Journal of Bacteriology* **155**, 734–741.

Smart, C. D., Sears, B. B. and Kirkpatrick, B. C. (1994). Analysis of evolutionary relationships between MLOs and other members of the class mollicutes based on 16S/23S rRNA intergenic sequences. *IOM Letters* **3**, 269–270.

Smith, A. J., McCoy, R. E. and Tsai, J. H. (1981). Maintenance *in vitro* of the aster yellows mycoplasma-like organism. *Phytopathology* **71**, 819–822.

Sundstrom, T. K. and Wieslander, A. (1990). Plasmid transformation and replica filter plating of *Acholeplasma laidlawii*. *FEMS Microbiology Letters* **72**, 147–152.

Swofford, D. L. (1993). PAUP: phylogenetic analysis using parsimony, version 3.1.1. Computer program distributed by the Illinois Natural History Survey, Champaign, Illinois.

Toth, K. F., Harrison, N. and Sears, B. B. (1994). Phylogenetic relationships among members of the class Mollicutes deduced from rps3 gene sequences. *International Journal of Systematic Bacteriology* **44**, 119–124.

Tully, J. G. (1989). Class Mollicutes: new perspectives from plant and arthropod studies. *In* "The Mycoplasmas", Vol. 5 (R. F. Whitcomb and J. G. Tully, eds), pp. 1–27. Academic Press, New York.

Tully, J. G., Rose, D. L., McCoy, R. E., Carle, P., Bove, J. M., Whitcomb, R. F. and Weisburg, W. (1990). *Mycoplasma melaleucae*, sp. nov., a sterol-requiring mollicute from flowers of several tropical plants. *International Journal of Systemic Bacteriology* **40**, 143–147.

Tully, J. G., Bove, J. M., Laigret, F. and Whitcomb, R. F. (1993). Revised taxonomy of the class Mollicutes: proposed elevation of a monophyletic cluster of arthropod-associated Mollicutes to ordinal rank (Entomoplasmatales ord. nov.), with provision for familial rank to separate species with nonhelical morphology (Entomoplasmasmataceae fam. nov.) from helical species (Spiroplasmataceae) and emended descriptions of the order Mycoplasmatales, family Mycoplasmataceae. *International Journal of Systematic Bacteriology* **43**, 378–385.

Vignault, J. C., Saillard, C., Tully, J. G., Mouches, C., Bove, J. M. and Bernhard, R. (1980). Attempts to culture and characterize mycoplasma-like organisms from apple proliferation affected fruit. *Review of Infectious Diseases* **4** (Suppl.), 248.

Weisberg, W. G., Tully, J. G., Rose, D. L., Petzel, J. ., Oyaizu, H., Mandelco, L., Sechrest, J., Lawerence, T. G., Van Etten, J., Maniloff, J. and Woese, C. R. (1989). A phylogenetic analysis of the mycoplasmas: basis for their classification. *Journal of Bacteriology* **171**, 6455–6467.

Whitcomb, R. F., Jensen, D. D. and Richardson, J. (1966). The infection of leafhoppers by western X-disease virus. I. Frequency of transmission after injection or acquisition feeding. *Virology* **28**, 448–453.

Whitcomb, R. F., Tully, J. G., Bove, J. M. and Saglio, P. (1973). Spiroplasmas and acholeplasmas: multiplication in insects. *Science* **182**, 1251–1253.

Williamson, D. L. and Whitcomb, R. F. (1975). A cultivable spiroplasma causes corn stunt disease. *Science* **188**, 1018–1020.

Woese, C. R. (1987). Bacterial evolution. *Microbiological Reviews* **51**, 221–271.

Woese, C. P, Stackebrandt, E. and Ludwig, W. (1985). What are mycoplasmas: the relationship of tempo and mode in bacterial evolution. *Journal of Molecular Evolution* **21**, 305–316.

Use of Categorical Information and Correspondence Analysis in Plant Disease Epidemiology

S. SAVARY

ORSTOM Visiting Scientist at IRRI, PO Box 933, 1099 Manila, Philippines

L. V. MADDEN

Department of Plant Pathology, Ohio Agricultural Research and Development Center, The Ohio State University, Wooster, Ohio 44691-4096, USA

J. C. ZADOKS

Department of Phytopathology, Wageningen Agricultural University, PO Box 8025, 6700EE Wageningen, The Netherlands

and H. W. KLEIN-GEBBINCK

IRRI, Department of Plant Pathology, PO Box 933, 1099 Manila, Philippines

Advances in Botanical Research Vol. 21
incorporating Advances in Plant Pathology
ISBN 0-12-005921-5

I. INTRODUCTION

It has long been recognized that a realistic approach to crop protection should consider the various pests, including diseases, that affect a crop (Padwick, 1956). Cropping practices represent major interactions within the disease tetrahedron (Zadoks and Schein, 1979) and, therefore, have to be considered when analyzing pathosystems. Systems analysis can provide an adequate set of concepts and tools to study the components of pathosystems and their interactions (Teng and Bowen, 1985). Since the number of components is large and their interactions are complex, a rationale is often needed to delineate the limits of the system to be addressed. A survey may provide the necessary overview of the pathosystem; adequate methods for analyzing survey data can produce preliminary information on its behaviour including major interactions. In this context, surveys can be considered as part of a systems approach.

Epidemiologists are frequently confronted with large data sets representing information on the characterization, the dynamics, or the behavior of a pathosystem. Examples are a survey data set, including information on disease intensity, cropping practices, and environmental data, and a germplasm database, where cultivars are represented by reaction types, disease intensities in field experiments, and quantitative measurements on the disease cycle. A range of methods has been developed, or adapted, by plant pathologists to analyze such data sets. The objectives of the analyses are as diverse as the methods, the first being to compact the information in such a way that interpretations can be made, hypotheses presented, and new experiments conducted.

A holistic description of a pathosystem should ideally involve characteristics as different as disease intensity over time and space, crop management, soil type and weather data, and socioeconomic information. Appropriate methods are needed by which the various facets of the available information can be explored. This paper presents an approach to mobilize and exploit information that is diverse in nature, precision, and accuracy.

We shall attempt to minimize the use of technical jargon. The reader needs only to know a few terms (Porkess, 1988): *class* (a collection of individuals, e.g. plots, cultivars, or sites, that share a common characteristic); *categorization* (the process of grouping individuals into classes according to numerical boundaries); *categorical analysis* (an analysis that addresses categories and/or qualitative data); *contingency table* (a table that shows a bivariate frequency distribution, using quantitative or qualitative classification). The term *correspondence analysis* will be introduced later in the text.

This paper is a practitioner's view of a set of statistical tools applied to phytopathology. Because these tools allow one to handle information in a fresh and encompassing way, their use by plant pathologists can offer new avenues for research. We have chosen to emphasize a particular, relatively simple analytical strategy, which is described in a hypothetical example. This strategy is then applied to a set of three very different, actual examples. These examples are addressed with dissimilar perspectives, and this may require some inflection in the details of the methodology. However, the set covers such a large range of issues that we feel this demonstrates the general value of the methodology. In a final section, an overview is offered where its application domain, and the perspectives of combinations with other techniques, are briefly discussed.

II. METHODOLOGY

A. QUALITATIVE AND QUANTITATIVE ATTRIBUTES OF PATHOSYSTEMS

Plant pathologists have been concerned primarily with quantifying epidemiological characteristics of pathosystems (Zadoks, 1978). This effort is illustrated by the attention paid to the quality of disease measurement (Large, 1966; James, 1974; Daamen, 1986a, b; Kranz, 1988; Campbell and Madden, 1990). The ideal disease assessment method should be both precise and accurate (Nutter *et al.*, 1991), as well as reproducible and unbiased. However, many characteristics of pathosystems, such as cultivars and the physiological races of the pathogen, are qualitative in nature. Other examples are the cropping season, the previous crop, or the soil type. Methods are therefore needed to consider these qualitative and quantitative attributes of pathosystems simultaneously.

A hypothetical germplasm database may, for instance, include the following information: field assessments using a 5-point grading scale based on disease severity and symptom pattern (IRRI, 1988, modified), disease reaction types (Chester, 1946; Zadoks and Schein, 1979; Savary *et al.*, 1989), with four categories, and quantitative measurements of the infection efficiency, the latent period, and the sporulation intensity (Table I). A plant pathologist may wish to analyze the relation of semiquantitative field assessments with quantitative measurements of monocyclic processes, and then check how the various categories of reaction types correspond to the monocyclic components and epidemic levels.

TABLE I

A hypothetical database on disease reaction of a germplasm collection: list of variables

Symbol	Variable[a]	Measurement	Unit
F	Field reaction	Rating given to a variety using a combination of quantitative and qualitative factors, in field experiments, using a grading scale, e.g.: 0: no disease 1: less than 1% severity (apical lesions) 2: 1–5% severity (apical and some marginal lesions) 3: 6–25% severity (apical and marginal lesions) 4: 51–100% severity (apical and marginal lesions)	None
I	Infection efficiency	Average proportion of deposited spores that produce lesions; in glasshouse experiments	Lesion/spore
L	Latent period	Average delay from inoculation to first sporulation in a population of lesions; in glasshouse experiments	Hours
S	Sporulation intensity	Average spore production in a population of lesions; in glasshouse experiments	Spore/lesion
R	Reaction type	Ranking of the varieties using a typological scale consisting of standardized classes characterizing a given level of host–pathogen compatibility, e.g.: — highly susceptible (HS) — susceptible (S) — resistant (R) — highly resistant (HR)	None

[a] In the example, data are assumed to be available for 78 cultivars.

B. CATEGORIZATION OF QUANTITATIVE INFORMATION

Statistical methods applicable to large, complex data sets usually imply a set of conditions, such as a linear relationship between the dependent and the independent variables for multiple linear regression analysis, and a series of prerequisites, among which is the homoscedasticity of the error term of the regression model (Butt and Royle, 1974; Teng and Gaunt, 1980; Campbell and Madden, 1990). Disease or pest variables usually do not comply with these prerequisites (McCool et al., 1986), and data transformation is often needed.

Another reason for transforming pest data is to represent mechanisms of damage they induce in the crop (Teng and Gaunt, 1980). Although appropriate transformation of data may, or may not (McCool et al., 1986), achieve these objectives, the result is often a reduction of the overall clarity of the model (Neter and Wassermann, 1974). Relatively complex transformations may, however, be considered in the case of simple systems, where a few independent variables are involved (Madden et al., 1981; Savary and Zadoks, 1992a).

Methods that allow the simultaneous handling of the two types of attributes — quantitative and qualitative — and that do not imply a priori assumptions on the variables, such as a linear relationship, are thus desirable. One way is to make quantitative variables compatible with qualitative variables, and encode them into classes, i.e. define quantitative boundaries of classes, and encode the values of the quantitative variables according to these boundaries. This encoding process allows the investigator: (a) to define the boundaries such that they represent the (maximum possible) error made in the measurement of each variable (variables with low accuracy would be represented by a few, broad classes, while variables with high accuracy would be represented by a larger number of classes); and (b) to link the definition of classes with key-values, thresholds, or any information that might be available beforehand. The process of converting quantitative data into coded data is flexible, different options being available depending on the variable at hand. There is no statistical restriction for this process. The further analysis of the resulting coded data by means of contingency tables and chi-square tests, however, depends on the class-filling, and therefore on the number of classes relative to the size of the sample.

In the above-mentioned example of a germplasm database, five grades (0 to 4) of disease intensity in the field are considered (Table I). When considering this variable, the analyst may take into consideration the facts that: (a) the assessments pertain to conventional designs, i.e., contiguous plots with a few border rows, resulting in interplot interferences; (b) small plots may not be representative of full-scale commercial fields; and (c) declaring complete absence of disease might have required a detailed inspection of every plant in each plot, a procedure incompatible with large varietal trials. These facts may have affected the disease assessment, especially at lower intensities. One may decide to merge grades 0 and 1 into one class, i.e., class 1, or "very low disease" and to consider three other classes, i.e. class 2 ("low"), class 3 ("medium"), and class 4 ("high disease"), representing the previous grades 2, 3, and 4, respectively. Classes similar in size are desirable. As the database involves 78 cultivars, it is expected that disease intensity in the field, when evenly distributed, would be represented by four classes, each containing approximately 20 cultivars. In this example, they contain 18, 21, 21, and 18 cultivars, respectively.

Infection efficiency (Table I) can be categorized in three classes: low ($I \leq 0.1$; I1), medium ($0.1 < I \leq 0.2$; I2) and high ($I > 0.2$ lesion/spore; I3).

TABLE II

A hypothetical database on a disease reaction of a germplasm collection: encoding of data[a]

	Initial data					Coded data				
Var.	L	S	I	F	R	L	S	I	F	R
1	6.8	3500	0.17	4	S	1	2	2	4	3
2	6.9	4200	0.08	3	R	1	2	1	3	2
3	6.8	1900	0.05	2	HR	1	1	1	2	1
4	6.9	1500	0.05	1	HR	1	1	1	1	1
5	7.1	3200	0.21	4	S	2	2	3	4	3
6	9.9	5100	0.22	3	VS	3	3	3	3	4
7	9.3	1400	0.21	2	S	3	1	3	2	3
.	.	.	.	.	.	.	.	.	.	.
.	.	.	.	.	.	.	.	.	.	.
.	.	.	.	.	.	.	.	.	.	.
78	9.5	2200	0.04	1	R	3	2	1	1	2

[a] For abbreviations, see Table I.

The three classes are defined in such a way that they are represented by similar frequencies: 24, 27, and 27 cultivars in class 1, 2, and 3, respectively. Three classes are also defined for the latent period (L) and the sporulation intensity (S) using 7 and 9 days as boundary values for L, and 2000 and 5000 spores/lesion as boundary values for S.

Reaction type, being a typical qualitative variable (it is categorical, and coded), here with four classes, is not categorized, the successive levels of resistance (HR to VS) being simply coded from 1 to 4.

The process of categorization and encoding is illustrated in Table II. Its main outcome is to translate information that was diverse in its format (figures, and numerical or alphanumerical codes) into a standardized, coded format.

C. CONTINGENCY TABLE AND CHI-SQUARE TESTS

Building contingency tables is an easy way to explore the relations between paired qualitative or coded quantitative variables of a data set represented by their successive classes. The contingency tables show bivariate frequency distributions. A chi-square test can be applied to confirm the suggested pattern of the data. The null hypothesis is the independence of the distribution frequencies of the two variables. The validity of the chi-square test relies on the expected sizes of groups; it should generally only be applied to groups with a minimum size. This minimum size was fixed at 5 by Dagnélie (1973). Gibbons (1976) gave a more flexible rule: no more than 20% of the expected

TABLE III

A hypothetical database on a disease reaction of a germplasm collection: a contingency table [a]

Latent	Field reaction [b]			
period	F1	F2	F3	F4
L1	3	2	7	15
L2	6	7	11	3
L3	9	12	3	0

[a] Numbers represent cultivars with a particular latent period and field reaction, chi-square value: 35.8; d.f. = 6; $P < 0.0001$.
[b] See Table I.

values (assuming independence) should be smaller than 5. In practice, major disequilibrium among the classes should be avoided, all classes being represented by commensurate numbers of individuals. This guideline is the only one that is essential to the next step, the analysis of a series of contingency tables by means of correspondence analysis. The way the analysis should be pursued is essentially dependent on the design of a strategy that adequately addresses the problem at hand.

In the germplasm database example, an important result of the categorization and coding of the data is a limited and consistent number (3 to 4) of classes for each variable, with classes being represented by approximately balanced numbers of cultivars. Simple contingency tables can, therefore, be assembled to analyse the relationships among variables, as, for instance, between disease intensity in the field (F) and latent period (L). The [F × L] contingency table (Table III) is a 4 × 3 matrix that shows that cultivars with very low disease (F1) usually have long latent periods (the profile is: three cultivars with short, six with medium, and nine with long latent periods), whereas cultivars with high disease (F4) usually have short latent periods (15 cultivars with short, three with medium, and none with long latent period). Profiles of relationships can also be examined row-wise: short latent periods (L1) are predominantly associated with high disease (three cultivars with very low, two with low, seven with high, and 15 with very high disease). The distribution frequencies of the two categorized attributes, F and L, are not independent as can be tested with a chi-square test, where the independence hypothesis is rejected ($P < 0.0001$).

Contingency tables can be assembled in the same way for all the variables of the database. Questions to be addressed include: (a) what is the relation between disease intensity in the field (F) and the monocyclic processes (I, L, and S); and (b) how well is R represented in terms of both disease intensity and monocyclic processes? The following strategy may be followed:

1. contingency tables are assembled using F classes as columns; in other words, disease intensity would be used as a "guide" throughout the analysis;

2. the three tables, [F × I], [F × L], and [F × S] are put together so that the table is of dimension 4 × 9; then

3. the relationships between these two types of variables, disease intensity and monocyclic components, are explored using correspondence analysis; finally,

4. the associations between F, I, L, S, and reaction types (R), are analysed using the contingency table [F × R] in the framework developed in the previous step.

D.   CORRESPONDENCE ANALYSIS: PROCEDURE

Correspondence analysis (Benzécri, 1973; Hill, 1974; Greenacre, 1984) is a multivariate statistical method to represent contingency tables in pictorial and tabular form. The data handled in the analysis are classes of coded variables, as, for example, the columns and rows of the contingency tables shown in Table IV. Each class is represented by its profile, either by rows (e.g. I1 is represented by its profile in terms of disease intensities: 12, 9, 3, 0) or by columns (F1 is represented by its profile in terms of monocyclic processes: 12,

TABLE IV

A hypothetical database on a disease reaction of a germplasm collection: contingency tables arranged for the two steps of a corresponding analysis[a]

	Field reaction			
Variable[b]	F1	F2	F3	F4
I1	12	9	3	0
I2	6	9	10	2
I3	0	3	8	16
L1	3	2	7	15
L2	6	7	11	3
L3	9	12	3	0
S1	12	9	3	0
S2	4	6	8	9
S3	2	6	10	9
R1	12	7	1	0
R2	3	0	11	0
R3	3	11	3	9
R4	0	3	6	9

[a] See Tables I and III.

[b] Variables in bold and italics are used in first and second steps of the analysis, respectively (see text for further explanation).

6, 0; 3, 6, 9; and 12, 4, 2, for I, L, and S, respectively). According to the strategy in the example, the vertical profiles of disease intensities in terms of reaction types are not yet considered; the upper part of Table IV, a 4 × 9 matrix, is first analysed. The associations of several variables belonging to two types (disease intensity in the field F, i.e., the result of a polycyclic process, and monocyclic attributes of the pathosystem: I, L, and S) are examined simultaneously. The procedure is similar to principal component analysis, and involves the computation of eigenvalues and eigenvectors (Benzécri, 1973; Greenacre, 1984). The sum of the eigenvalues is called the *inertia* and, with correspondence analysis, equals the chi-square statistic divided by the total number of observations. Each class contributes a fraction to the total inertia; summation of inertia over classes produces the total inertia. Coordinates for new axes are defined based on the eigenvalues, unlike principal component analysis, where the entries of the data matrix are quantitative, correspondence analysis entails computations on a data matrix of frequencies. Another difference is that classes in the columns and the rows are involved in the same way: each eigenvector is made up of weighted combinations of all the classes that have been selected for the analysis. The reason for this is that correspondence analysis is based on a chi-square distance d between classes, that can be written as (Dervin, 1988):

$$d(i, i') = \left(\sum_{j=1}^{p} (X_{ij}/X_{i.} - X_{i'j}/X_{i'.})^2/X_{.j} \right)^{1/2}$$

where: i and i' are two rows (or columns),
   X_{ij} is the frequency value in the ith row and jth column (e.g. $X_{23} = 10$),
   $X_{i.}$ is the sum of the frequencies along row (or column) i (e.g. $X_{2.} = 27$),
   $X_{.j}$ is the sum of the frequencies along column (or row) j.

This distance definition, which can be applied to rows or columns, differs from the Euclidian distance used in principal component analysis:

$$d(i, i') = \left(\sum_{j=1}^{p} (X_{ij} - X_{i'j})^2 \right)^{1/2}$$

where: i and i' are two rows (or columns, depending on the type of ordination used), representing two individuals (e.g., fields, cultivars, or epidemics, *not frequencies*),
   j is the index for variables (quantitative descriptors),
   X_{ij} is the measure of variable j representing the ith individual.

Details on the mathematics of correspondence analysis can be found in Benzécri (1973), Greenacre (1984), and Dervin, (1988).

Table V shows the result of correspondence analysis applied to the matrix

TABLE V
Correspondence analysis: relative weights and contributions to axes

		Contribution to axes					
		Axis 1			Axis 2		
		Contribution		Co-ordinate	Contribution		Co-ordinate
Classes	Relative Weight	To axis	Reciprocal		To axis	Reciprocal	
Columns							
F1	0.231	28.7	89.7	−0.69	20.5	7.9	−0.21
F2	0.269	13.7	88.6	−0.44	1.0	0.8	+0.04
F3	0.269	3.4	31.7	+0.22	56.6	64.9	+0.32
F4	0.231	54.2	95.2	+0.95	21.9	4.7	−0.21
Rows							
L1	0.115	17.6	89.0	+0.77	15.9	9.9	−0.26
L2	0.115	0.4	10.4	−0.12	25.3	78.0	+0.32
L3	0.103	14.2	89.5	−0.73	1.2	1.0	−0.08
S1	0.103	16.2	91.8	−0.78	10.3	7.2	−0.22
S2	0.115	2.6	99.4	+0.29	0.0	0.1	+0.01
S3	0.115	4.8	81.3	+0.40	8.4	17.4	+0.19
I1	0.103	16.2	91.8	−0.78	10.3	7.2	−0.22
I2	0.115	1.7	36.1	−0.24	24.8	63.9	+0.32
I3	0.115	26.2	97.7	+0.93	3.9	1.8	−0.13
Additional variables							
R1	0	-	79.2	−0.90	-	17.7	−0.43
R2	0	-	0.1	+0.04	-	58.8	+0.94
R3	0	-	6.8	+0.14	-	13.3	−0.20
R4	0	-	97.9	+0.77	-	0.1	+0.03
Inertia accounted for by axes		86.8%			10.7%		

of Table IV. As in principal component analysis, several axes are defined (equal to the smaller dimension of the table minus 1), but only the first two axes are given. Axes one and two accounted for 86.8 and 10.7%, respectively, of the total inertia of the data set. In general, with small contingency tables such as the one used in this example, more than two axes are seldom needed to account for more than 90% of the inertia. Besides the coordinates in the newly defined axes, the classes are represented by their relative weights, contribution to each axis, and reciprocal contribution to each axis. The relative weight (or mass) of each class represents the frequency of individuals in the corresponding row (or column). For instance, L1 = 0.115 [= (3 + 2 + 7 + 15)/(3 × 78)]. The contribution to an axis is the proportion (or percentage) of inertia of that *axis* which is derived from a specified class (17.6% for L1). The reciprocal contribution (or class correlation) represents the proportion of inertia of the class (row or column) that is accounted for by the specified axis. It is also the correlation between the axis and the class. For L1, 89% and 15.9% of the inertia of this class are accounted for by axes one and two, respectively (Table V). Finally, the sign of the new coordinate indicates the direction that the class deviates from the origin (i.e. marginal frequencies).

Graphs can be generated with correspondence analysis where (newly defined) coordinates of classes, rather than the frequencies, are plotted along the axes. The graph in Fig. 1A, where axis 1 is horizontal, and axis 2 vertical, illustrates the relations among classes. Proximity of points representing classes indicates correspondences that can be checked using chi-square tests. When a series of classes representing successive levels of a coded quantitative or of a semiquantitative variable that reflects a logical increase (as, for example, disease intensity, F1–F4) is considered, a path linking the successive classes can be drawn, and the movement along this path may be examined in relation with positions of other classes, paths, and axes. Fig. 1A shows the relationships among F, I, L, and S. High disease intensity (F4) is graphically associated with (close to) high infection efficiency (I3) and short latent period (L1); low disease intensity (F1) is associated with low infection efficiency (I1), long latent period (L3), and low sporulation (S1). The shapes of the paths of increasing disease intensity (F1–F4), increasing infection efficiency (I1–I3), and decreasing latent period (L3–Ll) are similar, indicating strong correspondences. The relation between disease intensity and sporulation intensity (S1–S3) is not as strong. The two paths, although sharing the same movement along the direction of the horizontal axis, and starting in the same area of the graph, diverge in their later part, and do not have the same length. The graph therefore suggests that, whereas low sporulation is closely associated with low disease intensity, medium (S2) to high (S3) sporulation corresponds to similar disease intensity levels, predominantly medium (F3) and high (F4); this interpretation of the graph adequately describes the corresponding contingency table (Table IV, [F × S]).

The next step of the analysis, the evaluation of representativeness of reaction types, is addressed in the second part of Table V (see "Additional variables"). In the case of this additional variable, no contribution to axes is computed, since it was not involved in the computation of eigenvalues and eigenvectors. What is actually done is that the axes defined in the first step of the analysis are used as a framework, on which the new variable, represented by its four classes is superimposed. Reciprocal contributions to axes are computed, which allows an assessment of how well the additional classes fit in with the framework. This is measured by the relative contribution values: $79.2 + 17.7 = 96.9\%$ of the inertia represented by R1 is accounted for by the two first axes. Similarly, 58.9, 20.1, and 98.0% of the inertias represented by R2, R3, and R4, respectively, are accounted for. It can, therefore, be concluded that R1, R2, and R4 are mostly represented by the two first axes, whereas the representation of R3 is poor. A further inspection of computation results would indicate that 79.9% of the inertia of this reaction type is accounted for by axis 3, which in turn accounts for only 2.4% of the total information represented by the upper part of Table IV.

The four reaction types can be plotted onto the previous graph (Fig. 1B). As expected, R1 is closely associated with low infection efficiency, long latent period, low sporulation, and low disease intensity; R4 is associated with high infection efficiency, short latent period, medium to high sporulation, and high disease intensity. R2 and R3 occupy intermediate positions.

The overall conclusion of the analysis of this hypothetical example would therefore be that a close association between disease intensity and monocyclic processes, especially infection efficiency and latency period, exists. Very susceptible (R4) and highly resistant (R1) reaction types adequately predict epidemic levels; but caution is needed when considering the epidemiological effect of R2 and, especially, R3.

III. EXAMPLES

A. EPIDEMIC TRENDS IN GROUNDNUT DISEASES

A survey was conducted in Ivory Coast on the main diseases of groundnut (Savary, 1987). In West Africa, several diseases affect groundnut crops, and the aim of the survey was to assess the part taken in this multiple pathosystem by a new component, groundnut rust (*Puccinia arachidis* Speg., Savary *et al.*, 1988). The survey was conducted during three consecutive years, yielding information on various components of the pathosystem, especially foliar diseases, in a population of 309 fields. The fields were visited at various stages in the crop development, and the analysis considered each field at each visit as a unique observation. One of the analyses performed on these data was aimed at obtaining a simplified, overall picture of a typical farmer's field, in terms of dynamics of the main diseases. Three disease variables were

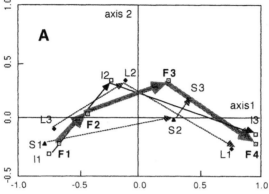

- ⊡ F: disease intensity
- ▫ I : infection efficiency
- ◆ L: latent period
- ▲ S : sporulation intensity

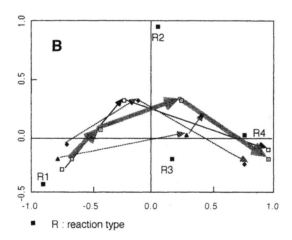

- ■ R : reaction type

Fig. 1. A correspondence analysis of a theoretical germplasm database on disease reaction. (A) First step of the analysis, involving disease intensity (F), infection efficiency (I), latent period (L), and sporulation intensity (S). The graph is drawn using the first (horizontal) and second (vertical) axes of the analysis. The successive classes are indicated and linked into paths. (B) Second step of the analysis. Four reaction types (R) were superimposed on the graph.

TABLE VI

Contingency tables and relationships among rust, late leaf spot and early leaf spot severities, and peanut development stages

Variables	Rust severity				
	R0	R1	R2	R3	R4
P0	53	8	1	0	0
P1	14	21	11	16	2
P2	8	13	3	13	1
P3	1	10	8	36	13
P4	2	2	9	32	33
A0	26	4	0	9	3
A1	33	22	9	19	6
A2	6	4	5	13	11
A3	9	12	6	36	20
A4	3	12	12	20	9
DV1	44	6	0	2	0
DV2	24	20	5	7	2
DV3	5	13	11	25	5
DV4	3	10	10	33	11
DV5	1	5	6	30	31

[a] R0–R4: classes of rust severities; P0–P4: classes of late leaf spot severities; A0–A4: classes of early leaf spot severities; *DV1–DV5*: crop development stages. Entries are number of fields. Each contingency table shows the bivariate frequency distribution of the sample (309 fields) for each pair of variables.

considered: rust, early leaf spot (*Cercospora arachidicola* Hori), and late leaf spot (*Phaeoisariopsis personata* (Berk. & Curt.) Deighton) severities. An additional variable, the crop development stage, was incorporated into the analysis in order to provide a physiological time reference. The frequency distributions of all three diseases were strongly asymmetrical, a large number of fields (especially those at early crop development) being unaffected, and a small number showing high severities.

Severity variables for the three diseases were encoded by use of the following boundaries: absent [0%]; very low [0, 0.01%]; low [0.01, 1%]; medium [1, 20%]; and high [20, 100%). For each disease, five severity classes were thus defined, labeled R0, R1, R2, R3, and R4 for rust, P0, P1, P2, P3, and P4 for late leaf spot, and A0, A1, A2, A3, and A4 for early leaf spot. Using these boundaries, we represented all classes by an adequate number of fields (30–100). Five development stages were considered: seedling to fourth tetrafoliate (DV1), flowering to beginning peg (DV2), beginning pod to beginning seed (DV3), full pod to full seed (DV4), and harvest maturity (DV5).

Three contingency tables were built (Table VI), by use of the the quantitative coded variables and the qualitative variable: [R × P], [R × A], and [R × DV]. Examination of the contingency tables provides indications on the relations among diseases, and between rust and physiological time. In the first contingency table, most of the fields were distributed along the first diagonal: there was an overall increase of late leaf spot with increasing rust severity. In the second table, most of the fields at low rust and early leaf spot levels were also concentrated along or adjacent to the first diagonal: R0 corresponded to A0 (26 fields) or A1 (33 fields); R1 corresponded to A1 (22 fields). But very high rust (R4) did not correspond to high early leaf spot (A4, 9 fields), but to medium early leaf spot (A3, 20 fields). The third table, [R × DV], represents the increase of rust with the increasing development of the crop: most of the fields were distributed, again, along the first diagonal. It is worth noting, however, that the highest frequency (mode) of rust severity at DV1 (44 fields), as well as at DV2 (24), was R0; at DV3, the mode reached R3 (25), where it remained at DV4 (33); and at DV5, the mode was at the highest rust severity, R5 (31 fields). This third table indicates that the increase of rust over development stages involves two lag phases, at epidemic onset, and at the end of the epidemic. All three tables can be submitted to chi-square tests (Table VI), where independence of the distributions of the four variables was rejected ($P < 0.0001$).

A correspondence analysis was conducted using the disease variables to generate a system of axes, and the development of the crop was superimposed on these axes. In other words, a framework representing interaction among diseases was first computed, onto which a scale of development stages was projected.

A series of axes was generated (Table VII), but only the first two axes are given, as they represented a total of 95.5% of the inertia of the analysed contingency table. The first axis represents a contrast between absence of rust (R0, in the negative direction, with a very strong contribution) and R3 and R4 (in the positive direction). It also represents a contrast between absence of late leaf spot (P0, in the negative direction) with P3 and P4. As for early leaf spot, it primarily accounts for an opposition between no or low disease severity (A0 and, to a lesser degree, A1), and other disease levels. Axis 1 essentially represents a contrast between unaffected or little diseased fields, and higher diseased fields. Axis 2 represents contrasts between low rust and late leaf spot levels (R1, P1, and P2), and high disease levels (R4 and P4); the contributions to this axis of early leaf spot (A) variables are small. As a summary, the first axis can be interpreted as representing appearance of diseases, and the second axis, intensification of diseases (at least, rust and late leaf spot) in the fields. The reciprocal contributions for the variables that were involved in axis generation are high; the proportion of inertia of R4 accounted for by the two first axes is, for instance: $64.5 + 33.0 = 97.5\%$. The large proportion of total inertia accounted for by axes reflects these large reciprocal contributions. It is worth noting that the reciprocal contributions for the categories of development stage (which were not involved in axis generation, and thus have no weight nor contributions to axes) were also large: the resulting axes account,

TABLE VII

Peanut disease severities: relative weights and contributions to axes

	Classes	Relative Weight	Axis 1 Contribution — To axis	Axis 1 Contribution — Reciprocal	Axis 1 Co-ordinate	Axis 2 Contribution — To axis	Axis 2 Contribution — Reciprocal	Axis 2 Co-ordinate
Columns	R0	0.249	64.3	96.3	−1.03	9.5	3.6	+0.20
	R1	0.175	0.7	5.3	−0.13	45.0	91.2	−0.52
	R2	0.104	2.8	36.1	+0.34	9.3	30.0	−0.31
	R3	0.314	14.3	87.2	+0.43	0.1	0.1	−0.02
	R4	0.159	17.9	64.5	+0.68	36.1	33.0	+0.49
Rows	P0	0.100	47.3	95.2	−1.39	9.0	4.6	+0.31
	P1	0.104	0.4	5.1	−0.12	26.4	90.1	−0.52
	P2	0.061	0.2	3.9	−0.11	13.0	77.5	−0.47
	P3	0.110	8.7	84.9	+0.57	0.6	1.5	−0.08
	P4	0.125	18.1	70.4	+0.77	27.8	27.6	+0.48
	A0	0.068	10.4	81.8	−0.79	7.0	14.0	+0.33
	A1	0.144	5.0	81.0	−0.37	4.1	16.9	−0.17
	A2	0.063	1.6	62.1	+0.33	2.9	28.3	+0.22
	A3	0.134	4.8	80.7	+0.38	1.5	6.6	+0.11
	A4	0.091	3.5	54.7	+0.40	7.2	28.9	−0.29
Additional variables	DV1	0	—	94.1	−1.36	—	5.8	+0.34
	DV2	0	—	65.6	−0.57	—	22.5	−0.33
	DV3	0	—	33.3	+0.29	—	52.7	−0.37
	DV4	0	—	79.7	+0.48	—	5.6	−0.13
	DV5	0	—	70.6	+0.77	—	25.4	+0.44
Inertia accounted for by axes				76.1%			19.4%	

Rust severity (R0–R4), and late (P0–P4) and early (A0–A4) leaf spot severity were involved in generation of axes. Development stage was used as an additional variable.

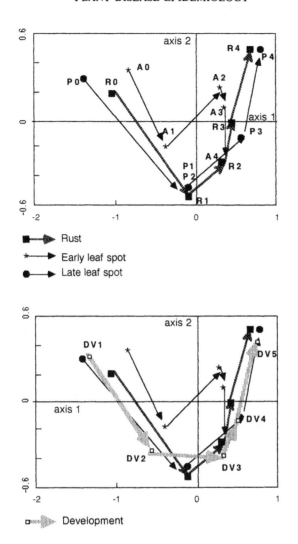

Fig. 2. A correspondence analysis of rust, late leaf spot, and early leaf spot of groundnut in farmer's fields in the Ivory Coast. Axes 1 and 2 are represented horizontally and vertically, respectively. The axes are defined using classes of rust (R), early (A) and late (P) leaf spot intensities. The successive development stages of the crop (DV) were superimposed on the graph.

to a very large extent, for the development of the crop.

Fig. 2 illustrates the associations indicated by the two axes. The graph indicates a very strong association between the paths of increasing rust and late leaf spot severities: the two paths show the same movement in the same direction. The association between the increase of rust and late leaf spot is also strongly corresponding to the development of the crop. The path representing early leaf spot initially corresponds to the paths of rust and late leaf spot (the movement from A0 to A1, and A2 parallels that of R0 to R3, and that of P0 to P3), but then shows a different movement (the increase from A2 to A3 and A4 corresponds to a direction opposite to the increase of rust from R3 to R4, and P3 to P4). This second phase of increase in early leaf spot intensity is also opposed in direction to the development of the crop (DV4 to DV5). As a result, maximum early leaf spot corresponds to beginning pod–beginning seed (DV3), not to harvest maturity. Therefore the graph indicates that the increase in rust and late leaf spot intensity corresponds to the overall increase in physiological age of the crop, whereas change in early leaf spot intensity consists of two phases: an initial increase, and a further decrease when the crop approaches maturity.

   Additional details in the description of variation in rust intensity over groundnut development are also indicated by the graph. R0 is close to DV1 (seedling to fourth tetrafoliate), and R1 is positioned between DV2 (flowering to first tetrafoliate) and DV3 (beginning pod to beginning seed); the delay to reach R1 is long. R2 is very close to DV3; the delay between R1 and R2 is, in contrast, short. R3 is close to DV4 (full pod to full seed), indicating an increased delay in disease increase from R2 to R3, which seems equivalent to the delay between R3 and R4. Based on this development scale, the progress of groundnut rust appears sigmoid: slow at the early crop development, then fast, and finally slower. The increase of late leaf spot is very similar except for the proximity of P1 and P2, suggesting an extremely fast disease increase in this phase of disease intensification. It is worth noting that A1 is very close to DV2, i.e., corresponds to an earlier development stage than P1 or R1. As a conclusion, this graph summarizes the appearance and further intensification of three groundnut diseases in an average farmer's field in Ivory Coast as follows: early leaf spot appears first, followed by rust and late leaf spot; the latter diseases have regular, sigmoidal progress throughout the cropping season, whereas early leaf spot reaches its maximum severity at V3 (beginning pod to beginning seed), after which severity decreases.

B. COMPONENTS OF RICE TUNGRO EPIDEMICS

Tungro is a major virus disease of rice in south and south-east Asia (Sogawa, 1976). The pathosystem involves several components: the spherical and bacilliform virus particles, vectors (among which *Nephotettix virescens* Distant plays a pre-eminent role), and the rice host plant (Ou, 1987). In some areas in southeast Asia, the disease appears to be present in every cropping season, whereas in some others, sporadic, and sometimes most damaging outbreaks occur. One major challenge in the understanding of rice tungro epidemics lies in the explanation of the polyetic processes which account for the endemicity of the disease in some areas, and for irregular, explosive epidemics in some others. Analyses on components of the tungro pathosystem were carried out from historical survey data conducted in the Philippines (Savary *et al.*, 1993).

The data were collected by the Philippines Department of Agriculture at several sites. In one province of the southern Philippines, North Cotabato, 130 fields were visited at the tillering phase during nine consecutive cropping seasons. The data included: cropping season (CS), planting date (PD), tungro incidence (IN), total vector population (V), and proportion of viruliferous vectors (VV). The disease is known to be endemic to the site. During the observation period, the disease was present in all years and cropping seasons, and had an overall disease prevalence (percentage of fields with infection) of 84.6%.

An analysis of these data was conducted to provide an overall description of the relationships between some components of the pathosystem, and assess the possibility of adequately characterizing tungro epidemics with a small set of variables that might be useful in tungro management.

Three quantitative variables, IN, V, and VV, were measured with low precision. Incidence (IN) was defined as the proportion of field area affected, and was assessed using an approximate grading scale with higher detail at low disease level (i.e., 0, 1, 3, 5, 10, 15, 20, 30, ... 100% field area affected). Disease incidence was encoded into disease classes: absence of disease (IN = 0: IN0), low incidence (0 < IN ≤ 2%: IN1) , average (2 < IN ≤ 5%: IN2), and high incidence (5 < IN ≤ 100%: IN3). Estimates of the vector population (V) were obtained by counting individuals caught in ten sweepnet strokes above the canopy. Total vector counts were encoded as: low (0 ≤ V ≤ 5: V1), medium (5 < V ≤ 10: V2), high (10 < V ≤ 15: V3), and very high (V > 15 vectors: V4). The proportion of viruliferous vectors (VV) was encoded using four classes: none (VV = 0: VV0), low (0 < VV ≤ 5: VV1), medium (5 < VV ≤ 15: VV2), and high (VV > 15% of the vectors inducing a positive reaction in the transmission test: VV3). The planting dates were classified from very early (PD1) to very late (PD5) in each cropping season, and the two cropping seasons were categorized as rainy (RS) or dry (DS).

Using these five categorized variables, we built four contingency tables: IN × V, IN × VV, IN × PD, and IN × CS. From these contingency tables, chi-square values were computed, which indicated a significant association ($P < 0.05$) between variables in all cases. For instance, the IN × PD contingency table indicated lower disease incidence in off-season plantings and the hypothesis of independence of the two variables was rejected (P < 0.01) on the basis of the chi-square value (58.7).

Tungro incidence may be seen as the outcome of the interactions among V, VV, PD, and CS. Thus, the four contingency tables were consolidated, using the classes of the variable to be explained, IN, as columns, and the different classes of the other variables as rows. This set of contingency tables was submitted to correspondence analysis.

Two main axes, accounting for 72.2 and 25.31% of total inertia, were found. The first (horizontal) axis opposes absence of tungro (IN0) to presence (IN1, IN2, and IN3), whereas the second (vertical) axis opposes low (IN1) to high (IN3) incidence. As in the previous example, the two axes may therefore be interpreted as representing disease appearance and intensification. The axes, of course, also incorporate contribution of the row-variables. On the first axis, absence of tungro is associated with absence of viruliferous vectors and off-season (PD1 and PD5) plantings. The second axis involves a contrast between cropping seasons, and a strong contribution of early–intermediate plantings, associated with a very high incidence of tungro.

Fig. 3 shows the overall association among variables: absence of tungro (IN0, on right hand-side) is associated with absence of viruliferous vector and off-season plantings (PD1 and PD5); low tungro (bottom) is associated with late plantings (PD4), medium to high population of vector (V2 and V3), low to medium proportion of viruliferous vectors (VV1 and VV2), and dry season (DS). High tungro incidence (top) is associated with early or intermediate plantings (PD2 and PD3), large population of vector (V4), very high proportion of viruliferous vectors (VV3), and rainy season (RS). The path of increasing tungro incidence corresponds to that of increasing proportion of viruliferous vectors in a stepwise pattern. Whereas the path of increasing tungro incidence can be projected on both axes with two phases, disease appearance (IN0–IN1, horizontal axis) and disease intensification (IN1–IN2–IN3, vertical axis), the variation in total vector population (V) is almost entirely associated with axis 2. This suggests that the increase in vector population is primarily associated

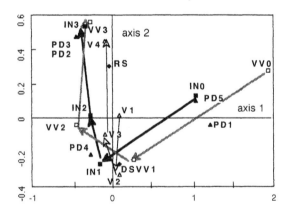

Fig. 3.   Characterization of rice tungro epidemics in one 'hot spot' in the Philippines. The graph represents data from a total of 130 fields visited during 9 consecutive cropping seasons at the tillering stage. Intensity of tungro (IN), total vector population (V), proportion of viruliferous vectors (VV), planting period (PD), and cropping season (RS or DS) were considered in the analysis. The classes that can be described by a natural increase (e.g. IN0 to IN3) were linked to show paths.

with disease intensification, not disease appearance. On the other hand, there is a close, stepwise correspondence between increasing proportion of viruliferous vectors, and tungro incidence. Hence, the analysis indicates that any increment of the proportion of viruliferous vectors is associated with a corresponding disease increase. This analysis was compared to those representing different endemic or non-endemic sites in the Philippines (Savary *et al.*, 1993), and was interpreted as a typical endemic situation.

C. RELATIONS BETWEEN PRODUCTION LEVEL AND YIELD LOSSES

A database for the pathosystem of groundnut was established from a series of six independent experiments (Savary and Zadoks, 1992a). Each experiment involved one main factor with three levels and three blocks. The main factors were water management, weed control, fertilizer input (two experiments), seeding rate, and cultivar potential yield. In addition, different disease treatments were used. Treatments were established within each block, representing combinations of manipulated levels of rust (low or high) and leafspot (low or high), with one control plot (C; no disease). The two diseases were independently manipulated by means of inoculations and sprays with a protectant fungicide. Each experiment included one reference block where inputs were set to default values, representing the production level of an average farmer's field: suboptimal water management and weed control, no fertilizer, low seeding rate, and cultivar with low potential yield. In each experiment, three types of yield were obtained. Actual yield (Y) was obtained from plots with different combinations of diseases and the main (input) factor. Attainable yield (Ya) was obtained from the control plots for each level of input factor. The attainable yield from the reference block was the reference yield (Yref) of an experiment.

Because the experiments were conducted over different seasons, and in a range of soil fertility, a large variation in reference yield was observed. The variation of actual yield was attributable to the superimposed effect of diseases. The overall relationships between Yref, Y, and presence or absence of diseases (D) was studied. The variable to be explained was Y, and two explanatory variables were considered: Yref and D. Only two categories of plots were therefore considered with respect to diseases: nearly disease-free, or affected by any of the two diseases. Variation of both yield variables was categorized in five classes, using the following boundaries: 0, 850, 1400, 2300, 3000, and 4500 kg/ha. Because plots used to estimate the reference yield were protected against diseases, no Yref values were found in the lowest class. Two contingency tables were built: Y × Yref and Y × D. Both indicated a significant association among variables ($P < 0.01$), and were submitted to correspondence analysis.

The two axes shown in Fig. 4 account for 62.6% (horizontal axis) and 21.8%

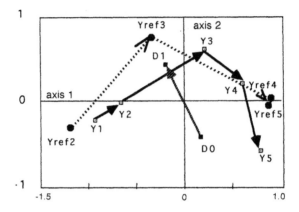

Fig. 4.   Overall relationship between reference yield, actual yield, and foliar diseases from an experimental database on groundnut. Reference yield (Yref) was measured in nearly disease-free plots, and represents a variation in production level due to climate and soil. Five classes of actual yield (Y) were considered; their boundaries are the same as for Yref, where no very low yield (Yref1) was recorded. Diseases are represented by two categories only: absent (D0) or present (D1).

(vertical axis) of inertia. The sequences of classes for each variable could be delineated (e.g., from Y1 to Y5), and the resulting paths analysed. This analysis indicated two phases in actual yield (Y) progress: (a) a first phase of parallel increase of the actual yield, along with the reference yield, independent of disease levels (the vector D0–D1 was orthogonal (perpendicular) to the Y direction), and (b) a second phase where the increase of actual yield, still related to that of reference yield, was opposed to the occurrence of diseases, i.e., the change in Y (Y3 to Y5) was opposite to the shift from D0 to D1. The graph indicates a threshold in reference yield, above which the increase of actual yield not only depends on the improvement of the production situation, but also on the protection of the crop against the two diseases. For instance, it was only when the reference yield was higher than Yref3 that the concurrent increase in actual yield was strongly influenced by (opposite to) the occurrence of diseases. In other words, the graph indicates a threshold in the improvement of the production situation, in terms of the hazard incurred by the two diseases. The threshold was represented by a class of reference yields ranging from 1400 to 2300 kg/ha (Yref3). It is worth noting that this represents the uppermost range of groundnut yields in farmers' fields in most of West Africa. The analysis indicates that progress in yield beyond this threshold cannot be foreseen without strict control of rust and leaf spot diseases. This analysis provides an overview of complex relationships in a two-disease pathosystem, that may be useful to set priorities in disease management. Further analysis of the data indicated that the two diseases differ in their harmful effects, and had less than additive effects on yield reduction (Savary and Zadoks, 1992c).

IV. OVERVIEW

A. COMBINATION AND LINK WITH OTHER TECHNIQUES

Many multivariate methods have been used by epidemiologists, and have been reviewed by Madden (1983), Hau and Kranz (1990), and James and McCulloch (1990). One area where these methods have been particularly useful is the analysis of survey data.

Examples of holistic pest surveys, which generated a better knowledge of multiple pests systems, include the analysis of bean constraints in Colombia (Pinstrup-Andersen et al., 1976), of yield-determining factors in pea in Idaho (Wiese, 1980), of wheat constraints in South Australia (Stynes, 1980) and in Germany (Kranz and Jörg, 1989). Numerous of methods were used to analyse the survey data, including regression analysis (Pinstrup-Andersen et al., 1976; Wiese, 1980), canonical correlation, principal component analysis (Stynes, 1980), and path coefficient analysis (Kranz and Jörg, 1989). Techniques involved in the comparison of epidemics are similar in the methodological approach (Kranz, 1974).

As mentioned earlier, correspondence analysis shares common features with principal component analysis on the computational aspects. Principal component analysis was used, for instance, to characterize tomato early blight epidemics (Madden and Pennypacker, 1979). The analysis yielded three major, independent, components representing maximum disease severity, rate of disease increase, and shape of the disease progress curves. Its result was to highlight, on the basis of a sample of epidemics, the value of key epidemiological attributes for characterizing epidemics.

Because it involves the use of qualitative information, correspondence analysis, as an analytical tool, can also be related to two other techniques, discriminant analysis and multiple regression. The main aim of discriminant analysis is to characterize separate groups in a population, by means of one or several discriminant functions (Madden et al., 1983; Hau and Kranz, 1990). Population densities of *Verticillium dahliae* and *Pratylenchus penetrans* at planting were, for instance, used to classify potato yields as being above or below 90% of the control (uninoculated) yield (Francl et al., 1987). Multiple regression involving qualitative ("dummy") variables (Chatterjee and Price, 1977) have seldom been used in plant pathology. Daamen (1986a) used this technique to analyse wheat mildew populations, and tested the effect of years, growth stages, and cultivars. Regression can also be used to evaluate the results of correspondence analyses. In the case of yield losses of groundnut, a backward, stepwise regression analysis yielded an equation of the following form (Savary and Zadoks, 1992a):

$$Y = a + b1\ Ya \times R1 + b2\ Ya \times S1 + b3\ R1 \times S1$$

where Y and Ya are the individual plot and attainable yields, respectively, and $R1$ and $S1$ are log-transformed areas under disease progress curves for rust and leaf spot, respectively. This equation describes yield variation as a response surface to interactions between attainable yield and disease injuries. Such a relationship is in agreement with the conclusion of the third example of correspondence analysis — the reference yield being the attainable yield associated with given climatic and soil conditions at a reference level of inputs. Further examination of the relationship indicated a less than additive effect of the two diseases on yield, which was confirmed by correspondence analysis on the same data set (Savary and Zadoks, 1992c), and independent factorial experiments (Savary and Zadoks, 1992b).

Another multivariate tool that can be associated with correspondence analysis is cluster analysis. Clusters can be considered as qualitative variables, among which relationships can be analysed. For instance, eight patterns of cropping practices, and eight pest (insects, diseases, and weeds) profiles were identified from the data resulting from a survey on rice crops in Central Luzon, Philippines (Elazegui et al., 1990), using two separate cluster analyses. These patterns of cropping practices and types of pest profiles were then related to categorized levels of yields (Savary et al., 1994).

B. USEFULNESS OF THE APPROACH

1. Precision of the data

Categorization of quantitative information representing components of a pathosystem is one means to take into account the level of precision of the data. This is particularly well exemplified in our second example on tungro epidemics. When numerous, precise data are available, a range of categories can be devised; when data with only low precision are available, a few, broad categories can be defined. This flexibility in handling data may be a critical advantage to enable the analysis of large, extremely valuable, sometimes heterogeneous data sets, such as those generated by plant protection or extension services.

2. Normalization

Categorization is also a means to normalize the variables (a) along their range, since each category should contain equivalent (commensurate) numbers of individuals (e.g., fields or epidemics), and (b) across data sets (e.g., surveys in different areas or years), so that comparisons are facilitated. Our third example on crop losses illustrates how a series of experiments, each of them representing a valuable fraction of the information needed for an overall interpretation, can be combined together without loss of vital information such as a variation in production situation, represented by the attainable yield.

3. Relations among variables

One strong advantage of dealing with categorized quantitative data is the rapid detection of complex, non-linear relations among several variables. This is best shown in our first example on groundnut diseases: all three disease variables are correlated with crop development, but with differing patterns, including lag phases, and disease decline. This can be detected with conventional non-linear regressions (Savary, 1987), at the cost of the choice of specific models. Correspondence analysis allows one to forward and test similar hypotheses, without specifying a functional relationship between development stage and disease intensity, thus providing an overall and neutral framework of relationships. These may further be explored more explicitly using regression techniques.

4. Thresholds

Finally, in the process of devising the boundaries and the categories, it is possible to introduce key or threshold values, the relevance of which might be revealed in the analyses. A domain where this approach might be particularly relevant is the development of strategies for tropical crop protection. Damage functions (Zadoks, 1985) are needed to define thresholds, which in turn are necessary to outline pest management strategies. These thresholds vary with production situations (De Wit and Penning de Vries, 1982) and are influenced by interactions between pests (Zadoks, 1985; Johnson et al., 1986; Savary and Zadoks, 1992a, b). An approach that consists of quantifying thresholds that are specific to each separate pest, in a range of production situations, will often be impracticable. Besides, yield estimates in the field are usually difficult, and involve high errors (Poate, 1988). So are yield loss estimates. Farmers often have a good perception of what the yield of their crop may be, or what the desirable yield of a crop is; they also have their own classification of yields, from very poor, over regular, to very good. Similarly, the farmer's input to the crop may be categorized as low, medium, or high. Such classifications are particularly relevant and convenient in the context of tropical small-scale farming. Decision-making in agriculture, either strategic (e.g., the choice of a cultivar), or tactical (e.g., to spray, not to spray, or to wait) is, by nature, categorical. Analogously, levels of hazard due to pests could be defined that would be conceptually compatible with decision-making.

C. THE NEED TO EXPLOIT CATEGORICAL INFORMATION IN PLANT PATHOLOGY

Correspondence analysis appears to be a robust technique that basically requires a conceptualization of a framework of relationships among components in the studied system. One might see it as just another multivariate tool where these components are represented in the form of classes, frequencies of classes, and then related in contingency tables. There are, however, solid reasons to consider correspondence analysis (and methods for analysing qualitative and categorical data in general) if plant pathologists are taking seriously the challenge of analysing the interactions that are at work within the disease tetrahedron. These interactions involve variables that are heterogeneous, and the approach outlined here is one means to analyze them.

In many pathosystems, patterns of cropping practices are driving variables of epidemics; whether a farmer's field belongs to a cluster representing intensive or extensive agriculture is a key to explain levels of a disease, or disease combinations (Savary et al., 1994). The pattern of cropping practice is, by essence, a qualitative attribute. Its introduction into analyses provides scope for a novel approach of the data, so that new hypotheses can be constructed.

REFERENCES

Benzécri, J. P. (1973). "L'Analyse des Données". Vol. 2. L'Analyse des Correspondances. Dunod, Paris.

Butt, D. J. and Royle, D. J. (1974). Multiple regression analysis in the epidemiology of plant disease. In "Epidemics of Plant Diseases" (J. Kranz, ed.), pp. 78-114. Springer Verlag, Berlin, Heidelberg, New York.

Campbell, L. C. and Madden, L. V. (1990). "Introduction to Plant Disease Epidemiology". John Wiley, New York.

Chatterjee, S. and Price, B. (1977). "Regression Analysis by Example", Chapter 4, pp. 74-100. Qualitative variables as regressors. John Wiley, New York.

Chester, K. S. (1946). "The Nature and Prevention of the Cereal Rusts as Exemplified in the Leaf Rust of Wheat". The Chronica Botanica Company, Waltham, MA.

Daamen, R. A. (1986a). Measures of disease intensity in powdery mildew (Erysiphe graminis) in winter wheat. 1. Errors in estimating pustule number. Netherlands Journal of Plant Pathology 92, 197-206.

Daamen, R. A. (1986b). Measures of disease intensity in powdery mildew (Erysiphe graminis) in winter wheat. 2. Relationships and errors of estimation of pustule number, incidence and severity. Netherlands Journal of Plant Pathology 92, 207-222.

Dagnélie, P. (1973). "Théorie et Méthodes Statistiques. Applications Agronomiques", Vol. 2. Presses Agronomiques de Gembloux.

Dervin, C. (1988). "Comment Interpréter les Résultats d'une Analyse Factorielle des Correspondances?" Institut Technique des Céréales et Fourrages, Paris.

De Wit, C. T. and Penning de Vries, W. W. T. (1982). L'analyse des systèmes de production primaire. In "La Productivité des Pâturages Sahéliens" (W. W. T. Penning de Vries and M. A. Djiteye, eds), pp. 275-283. Agricultural Research Report 918, Pudoc, Wageningen.

Elazegui, F. A., Soriano, J., Bandong, J., Estorninos, L., Jonson, I., Teng, P. S., Shepard, B. M., Litsinger, J. A., Moody, K. and Hibino, H. (1990). Methodology used in the IRRI integrated pest survey. In "Crop Loss Assessment in Rice", pp. 241-271. International Rice Research Institute, Los Baños, Philippines.

Francl, L. J., Madden, L. V., Rowe, R. C. and Riedel, R. M. (1987). Potato yield loss prediction and discrimination using preplant population densities of Verticillium dahliae and Pratylenchus penetrans. Phytopathology 77, 579-584.

Gibbons, J. D. (1976). "Non Parametric Methods for Quantitative Analysis". Holt, Rinehart and Winston, New York.

Greenacre, M. J. (1984). "Theory and Applications of Correspondence Analysis". Academic Press, London.

Hau, B. and Kranz, J. (1990). Mathematics and statistics for analysis in epidemiology. *In* "Epidemics of Plant Diseases. Mathematical Analysis and Modeling", 2nd edition. (J. Kranz, ed.), pp. 12–52. Springer Verlag, Berlin, Heidelberg, New York.

Hill, M. O. (1974). Correspondence analysis: a neglected multivariate analysis. *Applied Statistics* **23**, 340–354.

IRRI (International Rice Research Institute) (1988). "Standard Evaluation System for Rice", 3rd edition. International Rice Research Institute. Los Baños, Philippines.

James, F. C. and McCulloch, C. E. (1990). Multivariate analysis in ecology and systematics: panacea or Pandora's box? *Annual Review of Ecology Systematics* **21**, 129–166.

James, W. C. (1974). Assessment of plant diseases and losses. *Annual Review of Phytopathology* **12**, 27–48.

Johnson, K. B., Radcliffe, E. B. and Teng, P. S. (1986). Effects of interacting populations of *Alternaria solani, Verticillium dahliae* and the potato leafhopper *(Empoasca fabae)* on potato yield. *Phytopathology* **76**, 1046–1052.

Kranz, J. (1974). Comparison of epidemics. *Annual Review of Phytopathology* **12**, 355–374.

Kranz, J. (1988). Measuring plant disease. *In* "Experimental Techniques in Plant Disease Epidemiology" (J. Kranz and J. Rotem, eds), pp. 35–50. Springer Verlag, Berlin, Heidelberg, New York.

Kranz, J. and Jörg, E. (1989). The synecological approach in plant disease epidemiology. *Review of Tropical Plant Pathology* **6**, 27–38.

Large, E. C. (1966). Measuring plant disease. *Annual Review of Phytopathology* **11**, 47–57.

Madden, L. V. (1983). Measuring and modeling crop loss at the field level. *Phytopathology* **73**, 591–596.

Madden, L. V. and Pennypacker, S. P. (1979). Principal component analysis of tomato early blight epidemics. *Phytopathologische Zeitschrift* **98**, 364–369.

Madden, L. V., Pennypacker, S. P., Antle, C. E. and Kingsolver, C. H. (1981). A loss model for crops. *Phytopathology* **71**, 685–689.

Madden, L. V., Knoke, J. K. and Louie, R. (1983). Classification and prediction of maize dwarf mosaic intensity. *In* "Proceedings of an International Maize Virus Disease Colloquium and Workshop" D. T. Gordon, J. K. Knoke, L. R. Naut and R. M. Ritter, eds), pp. 238–242. The Ohio State University, Ohio Agricultural Research and Development Center, Wooster.

McCool, P. M., Younglove, T. and Musselman, R. C. (1986). Plant injury analysis: contingency tables as an alternative to analyses of invariance. *Plant Disease* **70**, 357–360.

Neter, J. and Wasserman, W. (1974). "Applied Linear Statistical Models". Richard D. Irwin, Homewood, Illinois.

Nutter, F. W., Teng, P. S. and Shokes, F. M. (1991). Disease assessment terms and concepts. *Plant Disease* **75**, 1187–1188.

Ou, S. H. (1987). "Rice Diseases", 2nd Edition. CAB International, Farnham House, Farnham Royal, Slough.

Padwick, G. W. (1956). Losses caused by plant diseases in the tropics. Commonwealth Mycological Institute Kew, Surrey, UK, Phytopathology Papers No. 1. 60pp.

Pinstrup-Andersen, P., de Londono, N. and Infante, M. (1976). A suggested procedure for estimating yield and production losses in crops. *PANS* **22**, 359–365.

Poate, D. (1988). A review of methods for measuring crop production from small-holder producers. *Experimental Agriculture* **24**, 1–14.

Porkess, R. (1988). "Dictionary of Statistics". Collins, London and Glasgow.

Savary, S. (1987). Enquête sur les maladies fongiques de l'arachide (*Arachis hypogaea*) en Côte d'Ivoire (I). Méthodes d'enquête et étude descriptive: les conditions culturales et les principales maladies. *Netherlands Journal of Plant Pathology* **93**, 167–188.

Savary, S. and Zadoks, J. C. (1992a). Analysis of crop loss in the multiple pathosystem groundnut-rust-leaf spot. I. Six experiments. *Crop Protection* **11**, 99–109.

Savary, S. and Zadoks, J. C. (1992b). Analysis of crop loss in the multiple pathosystem groundnut-rust-leaf spot. II. Study of the interaction between diseases and crop intensification in factorial designs. *Crop Protection* **11**, 110–120.

Savary, S. and Zadoks, J. C. (1992c). Analysis of crop loss in the multiple pathosystem groundnut-rust-late leaf spot. III. Correspondence analyses. *Crop Protection* **11**, 229–239.

Savary, S., Subba Rao, P. V. and Zadoks, J. C. (1989). A scale of reaction types of groundnut to *Puccinia arachidis* Speg. *Journal of Phytopathology* **124**, 259–266.

Savary, S., Noirot, M., Bosc, J. P. and Zadoks, J. C. (1988). Peanut rust in West Africa: a new component in a multiple pathosystem. *Plant Disease* **78**, 1001–1009.

Savary, S., Tiongco, E., Fabellar, N. and Teng, P. S. (1993). A characterization of rice tungro epidemics from historical survey data in the Philippines. *Plant Disease* **77**, 376–382.

Savary, S., Elazegui, F. A., Moody, K. and Teng, P. S. (1994). Characterization of rice cropping practices and multiple pest systems in the Philippines. *Agricultural Systems* **46**, 385–408.

Sogawa, K. (1976). Rice tungro virus and its vectors in tropical Asia. *Review of Plant Protection and Research* **9**, 21–46.

Stynes, B. A. (1980). Synoptic methodologies for crop loss assessment. *In* "Assessment of Losses Which Constrain Production and Crop Improvement in Agriculture and Forestry" (P. S. Teng and S. V. Krupa, eds), pp. 166–175. Miscellaneous Publication 7, University of Minnesota Agricultural Experiment Station.

Teng, P. S. and Bowen, K. L. (1985). Disease modelling and simulation. *In* "The Cereal Rusts", Vol. II. (A. P. Roelfs and W. R. Bushnell, eds), pp. 435–466. Academic Press, New York.

Teng, P. S. and Gaunt, R. E. (1980). Modelling systems of disease and yield loss in creals. *Agricultural Systems* **6**, 131–154.

Wiese, M. V. (1980). Comprehensive and systematic assessment of crop yield determinants. *In* "Assessment of Losses Which Constrain Production and Crop Improvement in Agriculture and Forestry" (P. S. Teng and S. V. Krupa, eds), pp. 262–269. Miscellaneous Publication 7, University of Minnesota Agricultural Experiment Station.

Zadoks, J. C. (1978). Methodology of epidemiological research. *In* "Plant Disease: An Advanced Treatise", Vol. II (J. G. Horsfall and E. B. Cowling, eds), pp. 63–96. Academic Press, New York.

Zadoks, J. C. (1985). On the conceptual basis of crop loss assessment: the threshold theory. *Annual Review of Phytopathology* **23**, 455–473.

Zadoks, J. C. and Schein, R. D. (1979). "Epidemiology and Plant Disease Management". Oxford University Press, New York.

AUTHOR INDEX

(Note: page numbers in italics refer to the full references)

SUBJECT INDEX